RECENT ADVANCES IN CHEMISTRY AND TECHNOLOGY OF FATS AND OILS

RECENT ADVANCES IN CHEMISTRY AND TECHNOLOGY OF FATS AND OILS

Edited by

R. J. HAMILTON and the late A. BHATI

Department of Chemistry and Biochemistry,
Liverpool Polytechnic, Liverpool, UK

ELSEVIER APPLIED SCIENCE
LONDON and NEW YORK

ELSEVIER APPLIED SCIENCE PUBLISHERS LTD
Crown House, Linton Road, Barking, Essex IG11 8JU, England

Sole Distributor in the USA and Canada
ELSEVIER SCIENCE PUBLISHING CO., INC.
52 Vanderbilt Avenue, New York, NY 10017, USA

WITH 69 TABLES AND 36 ILLUSTRATIONS

© ELSEVIER APPLIED SCIENCE PUBLISHERS LTD 1987

British Library Cataloguing in Publication Data

Recent advances in chemistry and technology
of fats and oils.
1. Oils and fats
I. Hamilton, R. J. II. Bhati, A.
664'.3   TP670

Library of Congress Cataloging in Publication Data

Recent advances in chemistry and technology of fats and
oils.

Includes bibliographies and index.
1. Oils and fats—Analysis. I. Hamilton, R. J.
(Richard John) II. Bhati, A. [DNLM: 1. Lipids—
analysis. 2. Oils—analysis. QU 85 R2945]
TP671.R43   1987   664'.3   87-600

ISBN 1-85166 070-4

The selection and presentation of material and the opinions expressed are the sole responsibility of the author(s) concerned

Special regulations for readers in the USA

This publication has been registered with the Copyright Clearance Center Inc. (CCC), Salem, Massachusetts. Information can be obtained from the CCC about conditions under which photocopies of parts of this publication may be made in the USA. All other copyright questions, including photocopying outside of the USA, should be referred to the publisher.

All rights reserved. No part of this publication may be reproduced, stored in a retrieval system, or transmitted in any form or by any means, electronic, mechanical, photocopying, recording, or otherwise, without the prior written permission of the publisher.

Photoset in Malta by Interprint Ltd
Printed in Great Britain by Page Bros. (Norwich) Limited

Preface

Since we produced *Fats and Oils: Chemistry and Technology* in 1980, the trend we anticipated to up-date the classical texts of oils and fats has manifested itself. Bailey's famous textbook has been completely revised and a second edition of Bernardini's work has been produced. The present text is an attempt to provide some insight into the current state of the art.

Chapter 1 discusses the physical properties of oils and fats with special reference to those properties which can be monitored to give an indication of the suitability of fats for chocolate production. The physical properties of the fats are often determined by the order in which the fatty acids are attached to the glyceride molecule. Ram Bhati, in the last article he wrote before his death, showed how mass spectrometry and chemical methods could be used to determine the sequence of fatty acids. Ram's essentially practical approach to the problem is exemplified by the section dealing with the experimental details of the techniques.

Chapter 3 outlines some of the problems which can arise in industry when the lipid part of a foodstuff undergoes oxidation, whilst in Chapter 4 Patterson describes the major technique, hydrogenation, which is used to circumvent the problems caused by oxidation of the unsaturated fatty acids. In Chapter 4 the essentials of the theory are given to enable the reader to appreciate the design features of the apparatus.

Chapter 5 deals with the analysis, mainly chromatographic, of lipids. Since milk fat is unique in commerce in having such a wide chain length range of fatty acids, it offers an opportunity to illustrate the special techniques needed to analyse these lipids.

In Chapter 6 Barnes provides a major review of the structure of the wheat grain, the lipid composition and the lipid distribution within the grain. From this essential description of lipids the role of these lipids in the baking of bread is explained.

The adulteration of commercial oils can be recognised by very accurate fatty acid analysis which is now possible by capillary gas liquid chromatography. However, the recognition of adulteration is dependent on having a standard fatty acid profile for one oil. Chapter 7 shows that there may be significant differences in the fatty acid composition from one variety to another within the same species. The review, though by no means exhaustive, attempts to bring together some results from the primary literature.

Podmore shows in Chapter 8 that the naturally occurring fats such as are described in Chapter 7 need to be modified before industry can use them. The three main methods of hydrogenation, interesterification and physical processing are exemplified by showing how soyabean, lard, margarine and palm oils are affected by them.

All of the chapters except one were originally presented at conferences and subsequently written up for this edition.

The editors wish to thank the Royal Society of Chemistry for permission to reproduce Dr Christie's article and the staff of Elsevier Applied Science for their help.

R. J. HAMILTON

Contents

Preface v
List of Contributors xi

1. Physical Properties of Fats and Oils 1
   P. J. M. W. L. BIRKER and F. B. PADLEY

   1.1 Introduction 1
   1.2 Melting and Crystallization Behaviour 2
   1.3 Solids Content and Phase Diagrams 6
   1.4 Compatibility of Fats: Product Defects 8
   1.5 Final Remarks 9
   1.6 References 10

2. Fatty Acid Sequence in Triglycerides and Related Compounds .. 13
   The late A. BHATI

   2.1 Introduction 13
   2.2 Methods for Determining Fatty Acid Sequence 15
   2.3 Selected Applications 22
   2.4 Appendix: Experimental Procedures 24
   2.5 References 29

3. Industrial Aspects of Lipid Oxidation 31
   J. C. ALLEN

   3.1 Introduction 31
   3.2 The Raw Materials 34
   3.3 The Finished Product 37

4. Hydrogenation of Oils and Fats 41
H. B. W. PATTERSON

4.1 Introduction 41
4.2 Layout 41
4.3 Autoclave Design 47
4.4 Reactions at the Catalyst Surface 52
4.5 References 56

5. The Analysis of Lipids with Special Reference to Milk Fat 57
WILLIAM W. CHRISTIE

5.1 Introduction 57
5.2 Lipid Class Separations 58
5.3 Fatty Acids and Related Aliphatic Compounds 61
5.4 The Positional Distributions of Fatty Acids in Milk Triacylglycerols 66
5.5 Molecular Species of Milk Triacylglycerols 68
5.6 Conclusions 73
5.7 Acknowledgement 74
5.8 References 74

6. Wheat Grain Lipids and their Role in the Bread-making Process   79
P. J. BARNES

6.1 Introduction 79
6.2 Structure of the Wheat Grain 79
6.3 Composition of Lipids in Wheat Grains 82
6.4 Relationship Between Grain Lipid Composition and Flour Lipid Composition 86
6.5 Storage Stability of Wheat Flour 95
6.6 The Role of Flour Lipids in Baking of Bread 96
6.7 Acknowledgements 105
6.8 References 106

7. Varietal Differences in Fatty Acid Compositions 109
R. J. HAMILTON

7.1 Introduction 109
7.2 Plant Lipid Compositions 110
7.3 Varieties Grown for Different Environmental Conditions .. 128
7.4 Compositions for Some Minor Seed Oils 141
7.5 References 166

8. Application of Modification Techniques.......................... 167
J. PODMORE

    8.1 Introduction 167
    8.2 Soybean Hydrogenation 170
    8.3 Lard — as a Shortening.................................. 172
    8.4 Margarine Development.................................. 172
    8.5 Palm Oil Utilisation....................................... 177
    8.6 Summary 180
    8.7 References 180

Index 183

List of Contributors

J. C. ALLEN
Research Division, North East Wales Institute, Deeside Industrial Park, Deeside, Clwyd CH5 2NU, Wales, UK.

P. J. BARNES
RHM Research Ltd, The Lord Rank Research Centre, Lincoln Road, High Wycombe, Bucks HP12 3QR, UK.

The late A. BHATI
Department of Chemistry and Biochemistry, Liverpool Polytechnic, Byrom Street, Liverpool L3 3AF, UK.

P. J. M. W. L. BIRKER
Unilever Research Laboratorium Vlaardingen, PO Box 114, 3130 AC Vlaardingen, The Netherlands.

WILLIAM W. CHRISTIE
Biochemistry Department, Hannah Research Institute, Ayr KA6 5HL, Scotland, UK.

R. J. HAMILTON
Department of Chemistry and Biochemistry, Liverpool Polytechnic, Byrom Street, Liverpool L3 3AF, UK.

F. B. PADLEY
Unilever Research, Colworth Laboratory, Colworth House, Sharnbrook, Bedford MK44 1LQ, UK.

H. B. W. PATTERSON
Consultant, 9 The Wiend, Bebington, Merseyside L63 7RG, UK.

J. PODMORE
Pura Foods Ltd, PO Box 27, Dunningsbridge Road, Liverpool L30 6XR, UK.

1

Physical Properties of Fats and Oils

P. J. M. W. L. BIRKER
*Unilever Research Laboratorium,
Vlaardingen, The Netherlands*
and
F. B. PADLEY
Unilever Research, Sharnbrook, UK

1.1 INTRODUCTION

The physical properties of fats and oils have been the subject of longstanding research efforts both in the academic world and in industrial research. Academic research has mainly been involved with fundamental studies often on pure triacylglycerols or simple mixtures of known composition. The edible fats industry, on the other hand, is often involved in the study of complex fat blends with the aim of tailoring their properties to suit specific applications in food products, and to control stability and shelf-life. Fractionation, hardening, interesterification, emulsification, crystallization and creaming are the most important processes for the production of a variety of fat based products such as margarines, halvarines, mayonnaise, creams, chocolate, speciality fats and table oils.

In many cases, the triacylglycerol composition of the fat blend used is not known in detail. Successful manufacture of these commodities relies on the manipulation of the fat blend in order to arrive at suitable physical properties, preventing undesirable changes in product properties during or after production. Well-known examples of such changes leading to product defects are the development of graininess in margarines and of bloom on chocolate.

An essential requirement for achieving control over production processes and product properties is a thorough understanding of the physical properties of oils and fats. Some important properties like melting and crystallization behaviour, polymorphism and crystal structure are discussed. Relevant physical analytical methods such as Differential

Scanning Calorimetry (DSC), X-ray diffraction (XRD), NMR, dilatometry and the use of phase diagrams in various forms are also described.

1.2 MELTING AND CRYSTALLIZATION BEHAVIOUR

1.2.1 Polymorphism

Melting and crystallization temperatures of triacylglycerols depend on two major factors: chemical structure and polymorphic behaviour.

The chemical composition of triacylglycerol mixtures can be modified by blending, interesterification, fractionation and by partial or complete hydrogenation. The way in which crystallization is controlled in the subsequent processes needed to manufacture products like margarines or chocolate largely determines the ultimate product properties. One of the most striking features of triacylglycerol crystallization is the occurrence of a variety of crystal modifications (polymorphism). Polymorphism is in fact deliberately used in the manufacture of fat products as one of the ways of controlling crystal shape and size and crystal–crystal interactions in products.

After a great deal of controversy in the early literature on fat polymorphism it is now generally accepted that triacylglycerols can crystallize in four major types of modification, known as sub-α, α, β' and β.

The β modification is the most stable one for most pure triacylglycerols. It is the only crystal form which has been characterized in exact detail by means of single crystal X-ray structure determination[1] (Fig. 1.1). Some natural fats of simple composition like cocoabutter can crystallize in the β form, depending on the experimental conditions.[2]

The β' modification is the stable solid form of certain types of triacylglycerol, such as those of the C_n C_{n+2} C_n-type, and also of most complex natural fats or fat blends. Examples of these are palm kernel, coconut or hardened cottonseed oils. The structure of triacylglycerols in the β' crystal modification is not known in great detail. A single crystal X-ray diffraction study of β'-triundecanoin has revealed some clues about the packing of the molecules in the crystal.[3] The direction of alkyl chains in the unit cell was found, but the glycerol residues could not be located. It is, moreover, quite likely that there is a great deal of structural variation within the β' class. Recent work at Unilever Research Vlaardingen has shown that twinned crystals of $\beta' - C_{12} C_{14} C_{12}$ crystallize in a nearly orthorhombic unit cell, with dimensions of $66.7 \times 5.7 \times 22.9$ Å, containing 8 molecules.

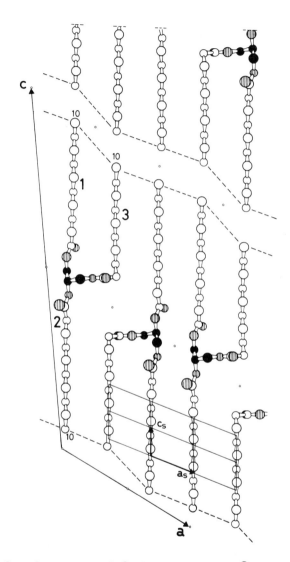

FIG. 1.1. Crystal structure of β-tridecanoylglycerol (● = glycerol carbon; ○ = chain carbon; ⦿ = oxygen). Small open circles between the molecules indicate the positions of inversion centres (from de Jong, Thesis, State University, Utrecht, 1980).

Systematic absences suggest the crystallographic space group Iba2 but the symmetry must in fact be somewhat lower as not all angles of the unit cell are exactly 90°. Further structural studies are in progress.

The α and sub-α modifications are always unstable. The sub-α ⟷ α phase transition is reversible, but transitions from α to β' or β are monotropic.[4]

1.2.2 The Study of Crystallization Processes

The study of fat crystallization naturally involves the measurement of crystal polymorphism and solids content in a fat blend. Temperature controlled X-ray diffraction and Differential Scanning Calorimetry (DSC) are the methods of choice for the study of fat polymorphism. DSC curves reveal transition temperatures and heats of fusion or crystallization, except for the reversible sub-α ⟷ α transition which is a second order (order-disorder) transition (Fig. 1.2a). DSC does not produce unequivocal information about the crystal modifications involved, especially when mixtures of triacylglycerol are being studied. The complete picture of triacylglycerol phase behaviour can only be obtained in combination with the complementary technique of temperature con-

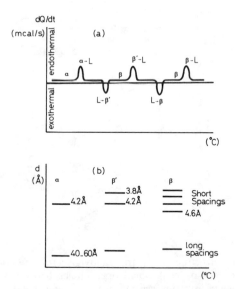

FIG. 1.2. Comparison of the information obtained from DSC(a) and X-ray diffraction (b) studies of triacylglycerols.

trolled X-ray diffraction (Fig. 1.2b). The situation is, however, not always as clear and well-resolved as shown schematically in Fig. 1.2. Peaks in DSC thermograms often show overlap, depending on the scan rate, and short-lived intermediates may be overlooked, especially in X-ray diffraction experiments where long exposure times are required.

The molar heats of fusion (ΔH_f) of triacylglycerols in different modifications can in principle be calculated from the integral surface under peaks in the DSC thermograms, if they are well separated, but this is not always the case.

It has been found that certain rules can be applied for the calculation of the molar heats of fusion. Those of saturated mono-acid β-triacylglycerols can, for instance be described by the following formula,[5]

$$\Delta H_f = 1.023 \times \text{(Carbon number)} - 7.79 \,(\text{kcal/mole}).$$

For partial triacylglycerols and triacylglycerols with unsaturated alkyl chains corrections have been worked out.[5]

DSC and X-ray diffraction can be applied to the study of pure triacylglycerols as well as for triacylglycerol mixtures, natural fats and fat blends, which mostly do not crystallize completely in temperature ranges of practical interest. In such cases it is important to know the extent of crystallization at various temperatures. Methods for measuring the solids content of fats are based on volume contraction during crystallization (dilatometry) and on differences in molecular mobility in liquid and solid triacylglycerols (wide-line NMR).

Pulse-NMR has replaced the laborious dilatometric analyses in most situations (e.g. Ref. 6), but much of the older literature refers to dilatation values.

The dilatation at a certain temperature (D_t) is related to the difference in volume between the completely liquid fat and the stabilized crystallized fat at that temperature. It is usually expressed in $mm^3/25\,g$ of fat. There is no unequivocal relationship between dilatation value and percentage of solids as the volumetric contraction is a function of molecular weight and crystal modification. Complete solidification corresponds with D_t values in the range of 2100–2500. The so-called solid fat index (SFI) has often been used to indicate the degree of crystallization. It is calculated as $D_t/25$. The SFI for a completely solid fat is consequently close to 100 in many but not all cases. The SFI measurement should, therefore, not be confused with the real percentage of solids as measured, for instance, by pulse-NMR.

1.3 SOLIDS CONTENT AND PHASE DIAGRAMS

The phase behaviour of a fat as a function of temperature or composition can be represented in a variety of phase diagrams. Figure 1.3 shows how dilatation or pulse-NMR data can be used to plot solids content as a function of temperature or composition. An alternative[6,7] and often convenient way of representing dilatation or solids content data as a function of composition is shown in Fig. 1.3 as well (iso-dilatation and iso-solids diagrams), where lines of constant solids content are plotted in a diagram of temperature versus composition.[6]

Iso-solids and iso-dilatation diagrams are well suited to show how two

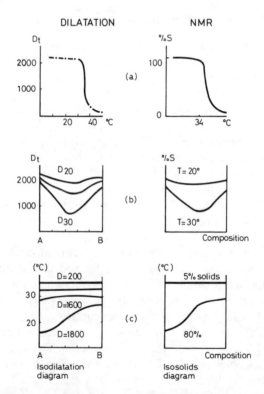

FIG. 1.3. Schematic representation of phase diagrams based on dilatation values and solids data from NMR experiments; (a) solids as a function of temperature for one blend; (b) solids as a function of composition and temperature; (c) iso-dilatation and iso-solids diagrams of compatible fats (see text).

triacylglycerols or two fats or fat blends interact. Well compatible fats have horizontal sets of iso-solids lines.

In other cases minima (eutectic behaviour) or maxima (compound formation) are observed.[6,7] Diagrams such as these are more useful for the description of phase behaviour than the conventional binary phase diagrams,[8,9] describing phase behaviour with solidus and liquidus lines in a diagram of temperature versus composition (Fig. 1.4). These phase diagrams should by definition represent thermodynamic equilibrium situations and can therefore in practice only be used for combinations of pure triacylglycerols in stable modifications. Figure 1.4 shows the serious effects of sample impurity and stabilization procedures on the quality of the resulting phase diagrams, even in such simple systems as $C_{16}C_{16}C_{16}/C_{18}C_{18}C_{18}$.[7] The mutual liquid solubilities of triacylglycerols—and thus the liquidus lines in these binary phase diagrams—are usually well described by the Hildebrand equation for ideal solutions,

$$\ln x = \frac{\Delta H_f}{R}\left(\frac{1}{T_m} - \frac{1}{T}\right),$$

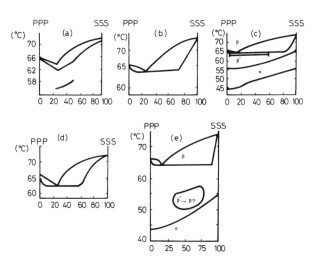

FIG. 1.4. Binary phase diagrams for the SSS/PPP system published since 1928, showing the effects of experimental technique (and possibly of impurities) on the shape of the solidus and liquidus lines; (a) based on observation of melting and solidification points (Ref. 11); (b) thaw–melt stabilization (Ref. 12); (c) diagram based on melting points, X-ray diffraction and dilatation values (Ref. 13); (d) thaw–melt technique (Ref. 8); (e) micro-DTA (Ref. 8) (figures adapted from Ref. 8).

or by its modified form[9] in the presence of solid solution (mixed crystals)

$$x_s \ln \frac{x_l}{x_s} + (1-x_s) \ln \frac{1-x_l}{1-x_s} = \frac{\Delta H_f}{R} \left(\frac{1}{T_m} - \frac{1}{T} \right)$$

where x_s and x_l are molar fractions in solid and liquid respectively, ΔH_f is the melting heat of the solid, R is the gas constant, and T_m is the melting point of the solid. The intersection of two Hildebrand solubility curves in the phase diagram corresponds with the eutectic point.

1.4 COMPATIBILITY OF FATS: PRODUCT DEFECTS

The physical properties of natural fats can be modified by hydrogenation, fractionation or interesterification to tailor them for use in food products. It is moreover often necessary to blend fats of different sources in order to obtain products with the desired melting point, solids content, consistency, etc. The physical properties of the blend lie ideally somewhere between those of the constituents at all ratios of mixing. Such combinations are said to be compatible, as for example in mixtures of cocoabutter (CB) and a good cocoabutter equivalent (CBE) like 'Coberine'[6] which is rich in 1,3-distearoyl-2-oleylglycerol (StOSt), 1-stearoyl-2-oleyl-3-palmitoylglycerol (StOP), and 1,3-dipalmitoyl-2-oleylglycerol (POP). The iso-solids diagrams of such blends contain approximately straight iso-solids lines (Fig. 1.5a). Other combinations like CB with palm kernel stearine ('laurics') are much less compatible (Fig. 1.5b). The crystal modification of the stabilized mixtures changes from β to β' when more than about 10% of laurics is added to CB.[6] At such concentrations the blend is very prone to post-crystallization of separate β crystals leading to a product defect known as bloom.[6] Combinations of CB with hydrogenated cottonseed oil fractions rich in saturated C_{16}, C_{18} and elaidic residues are entirely incompatible (Ref. 6, Fig. 1.5c).

A different example of a product defect due to incompatibility is the development of graininess in margarines based on blends containing, for instance, palm oil (with 30% of POP) and lard (with 50% of 2,3-dioleyl-2-palmitoylglycerol (OPO)). A phase study[10] has shown that this binary combination is not eutectic but forms a compound which is less soluble than either of the components. A 1:1 palm oil/lard mixture contains 28% solids at 22°C against 20 and 22% in pure palm oil and lard respectively.

The formation of another well-known compound, 1,3-dipalmitoyl-2-

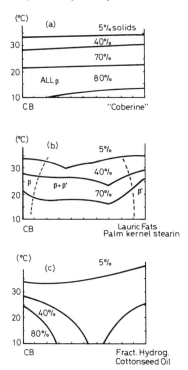

FIG. 1.5. Compatibility of cocoabutter with other fats as shown by iso-solids studies (adapted from Refs 6 and 7).

oleylglycerol/1,2-dipalmitoyl-3-oleylglycerol (POP/PPO), has been reported to result in maxima in the iso-dilatation diagrams[7] of an interesterified palm oil fraction (66% PPO) and palm mid-fraction (POP) at a 1:1 POP/PPO ratio (Fig. 1.6). Also see a more detailed recent review.[14]

1.5 FINAL REMARKS

Crystallization is always initiated by a nucleation process in an undercooled or supersaturated non-equilibrium system. Cooling rate, agitation, and the degree of undercooling are important parameters determining the relative rates of nucleation and crystal growth, and thus of crystal size and crystal agglomeration. Recent publications,[3,15] describe

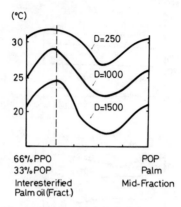

FIG. 1.6. Compound formation in a blend of POP and PPO-containing fats. The maxima in the iso-dilatation curves are found at the 1:1 POP/PPO ratio (from Ref. 7).

liquid fats as ordered systems in which the aligned long-chain triacylglycerol molecules form liquid crystalline structures of lamellar geometry. The presence of such phases would have a promoting effect on nucleation processes, especially for the formation of α-crystals, which are structurally related to a rigid lamellar phase. The existence of an ordered structure in the liquid phase would explain phenomena such as memory effects, observed when fats are melted and then crystallized. This area must still be regarded as poorly characterized.

Phase transitions into or crystallization of β'- or β-triacylglycerols result in crystals of different size and composition which play a functional role in the textural and melting properties of fat products, which can thus be tailored by blend formulation and processing.

1.6 REFERENCES

1. VAND, V. and BELL, J. P. (1951). *Acta Cryst.*, **4**, 465; LARSSON, K. (1964). *Arkiv Kemi*, **23**, 1; JENSEN, L. H. and MABIS, A. J. (1966). *Acta Cryst.*, **21**, 770; DOYNE, T. H. and GORDON, J. T. (1968). *J.A.O.C.S.*, **45**,333.
2. WILLE, R. L. and LUTTON, E. S. (1966). *J.A.O.C.S.*, **43**, 491.
3. HERNQVIST, L. and LARSSON, K. (1982). *Fette Seifen Anstrichmittel*, **84**, 349.
4. CHAPMAN, D. (1962). *Chem. Rev.*, **62**, 433.
5. TIMMS, R. E. (1978). *Chem. Phys. Lipids*, **21**, 113.
6. GORDON, M. H., PADLEY, F. B. and TIMMS, R. E. (1979). *Fette Seifen Anstrichmittel*, **81**, 116.

7. ROSSELL, J. B. (1973). *Chem. Ind.*, 832.
8. ROSSELL, J. B. (1967). *Adv. Lip. Res.*, **5**, 353.
9. KNOESTER, M., DE BRUYNE, P. and VAN DEN TEMPEL, M. (1972). *Chem. Phys. Lipids*, **9**, 309.
10. MORAN, D. P. J. (1963). *J. Appl. Chem.*, **13**, 91.
11. JOGLEKAR, R. B. and WATSON, H. E. (1928). *J. Ind. Chem. Soc.*, **47**, 365T; (1930). *J. Ind. Inst. Sci.*, **A13**, 119.
12. KERRIDGE, R. (1952). *J. Chem. Soc.*, 4577.
13. LUTTON, E. S. (1955). *J.A.O.C.S.*, **73**, 5595.
14. TIMMS, R. E. (1984). In *Progress in Lipid Research*. Ed. R. T. Holman, Pergamon Press, New York, U.S.A., Oxford, England, pp. 1–38.
15. LARSSON, K. (1972). *Fette Seifen Anstrichmittel*, **74**, 136.

2

Fatty Acid Sequence in Triglycerides and Related Compounds

The late A. BHATI
*Department of Chemistry and Biochemistry,
Liverpool Polytechnic, UK*

2.1 INTRODUCTION

Compared to other natural products such as alkaloids, proteins, steroids, terpenes, vitamins, etc., triglycerides and related phospholipids are simple compounds—just esters of glycerol with fatty acids, or fatty acids and phosphoric acids. Yet the determination of their structure and stereochemistry has not been so easy.

Most of the difficulties have been due to the lack of suitable methods of isolating pure compounds, as well as the nature of the triglyceride molecule. Apart from a few simple triglycerides, such as trimyristin, which occurs[1] alone in nut-meg oil, other triglycerides in other vegetable oils and animal fats occur as mixtures, from which isolation of pure individual triglyceride by conventional methods has been difficult, and even impossible.

The glycerol molecule, with its three hydroxyl groups, on esterification with three fatty acids can lead to a complex mixture of simple and mixed triglycerides. If the fatty acids involved are not very different in so far as the length and degree of unsaturation of their chains are concerned then the resultant triglycerides would have very similar physical properties, e.g. m.p., b.p., solubility, etc., and consequently be inseparable by conventional methods. Actually the major bulk of fatty acids of triglycerides of seed oils and animal fats comprise C_{16} and C_{18} acids with a smaller proportion of acids of shorter or longer chain lengths.

In spite of the above difficulties, the structures of unknown triglycerides were determined by the traditional chemists. Comparison with possible synthetic compounds provided confirmation of the structure.

A mixed triglyceride (I or II) in which the acyl groups attached to the primary oxygens of glycerol moeity are different, is capable of existing in two stereoisomeric forms:

$$\begin{array}{cc} CH_2OCOR^1 & CH_2OCOR^3 \\ | & | \\ CHOCOR^2 & CHOCOR^1 \\ | & | \\ CH_2OCOR^3 & CH_2OCOR^2 \\ (I) & (II) \end{array}$$

Although chemical methods are available for correlating the configuration of a new triglyceride with that of an authentic compound, a final check of their identity cannot be made with certainty by the ordinary polarimetric methods because the optical rotations involved are extremely low.

Nowadays powerful methods, such as thin layer chromatography (TLC), high performance liquid chromatography (HPLC), and gas chromatography (GC) are available for the isolation of pure triglycerides. In fact the combined application of gas chromatography – mass spectrometry (GC–MS) enables the direct determination[2,3] of structure of an individual triglyceride from a mixture.

The stereochemical problem of triglycerides has been solved in a novel way by employing both chemical and enzymic methods. In order to understand fully this new approach it is necessary to consider the prochirality of glycerol, and the stereochemical restrictions this imposes on the fatty esters derived. Glycerol has a *meso* carbon, Ca_2bd; and as with other compounds, such as ethanol, citric acid, etc., possessing *meso* carbons, its two similar groups whether free or in esterified form, can be distinguished by enzymes. It is, therefore, imperative, when referring to a mixed triglyceride or a symmetrical di-acid-diglyceride, to specify clearly which of the CH_2OCOR groups is involved in an enzymic reaction. None of the older methods, e.g. α, α', D,L and R,S systems of nomenclature is found[4] satisfactory for this purpose. The 'stereospecific numbering' (abbreviated *sn*) convention, proposed by Hirschmann[5] is now used. It recognises the fact that the two primary groups of glycerol, or its fatty ester are not identical in their reactions with disymmetric molecules such as enzymes, and if numbering of glycerol molecule is done with reference to the following projections, then

Fatty Acid Sequence in Triglycerides and Related Compounds

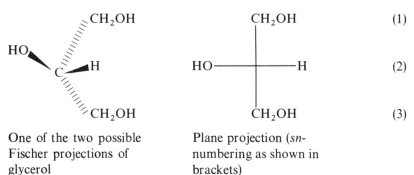

One of the two possible Fischer projections of glycerol

Plane projection (*sn*-numbering as shown in brackets)

not only can one clearly indicate which CH_2OCOR group is involved in enzyme specificity but also unambiguously express the stereochemistry of glycerol derivatives. The *sn* system does not have symbols to describe antipodal relationship as do the D, L or R, S systems but it makes it easy to understand and express configurational relationships of glycerides.

The fatty acid sequence in triglycerides and phospholipids represents the regio- and stereospecific location of the acyl groups on the glycerol back-bone (cf. plane projection formula above), i.e. which acyl group is present in the *sn* 1-position, and which in the *sn* 2-position, and so on.

There are academic, commercial and medical reasons for knowing the fatty acid sequence. The configuration of a given triglyceride or phospholipid can be unambiguously inferred from its fatty acid sequence. Allegedly similar products, e.g. 'butter and margarine' can be differentiated if the fatty acid sequence of constituent triglycerides are compared. A knowledge of the fatty acid sequence can be useful in metabolic studies, e.g. (i) which acyl group in which position of a triglyceride is easily and conducively metabolised; (ii) whether vegetable oil is consumed better without heating (as in salad-dressing) than on heating to a high temperature (as in frying). In the latter case, besides aerial oxidation, polymerisation, etc., there is the possibility of acyl migration and consequent alteration of fatty acid sequence.

The aim of this chapter is to review all the available methods for determining fatty acid sequence, along with some selected applications. An appendix describing experimental procedures is also given.

2.2 METHODS FOR DETERMINING FATTY ACID SEQUENCE

The fatty acid sequence determination can be divided into two parts. (A) Determination of acyl groups, attached to the two primary carbons, 1

(1)

(2) R^2CH_2COO ──┼── H

(3)

$$\begin{array}{c} CH_2OCOCH_2R^1 \\ | \\ \phantom{R^2CH_2COO\text{──}}| \\ | \\ CH_2OCOCH_2R^3 \end{array}$$
(III)

and 3; and the acyl group attached to the secondary carbon, 2 of the triglyceride (III). (B) Stereospecific distinction between the acyl groups attached to carbons 1 and 3.

A. Methods for finding out acyl groups attached to carbons 1 and 3
(a) Physical methods

On mass spectrometry at 70 eV (source temp. 170–250°) triglyceride (cf. III) ejects[6,7,2] acyloxymethylene ions $(CH_2OCOR^1)^+$, $(CH_2OCOR^3)^+$, indicating which groups are attached at the 1 and 3 positions. Full fledged development of this observation[6] into a method does not appear to have been made because the authors had also noticed a weaker $(CH_2OCOR^2)^+$ ion which obviously arises from acyl migration of COR^2 group from the 2- to the 1-position. Later workers[7] have noticed that ejection of ions $(CH_2OCOR^1)^+$, (CH_2OCOR^3) is a general behaviour of triglyceride (III). The acyl migration can be avoided by using milder mass spectrometric conditions, e.g. lower electron energy and specially lower ion-source temperature, or even using a softer mode of ionisation, such as field-desorption, fast atom bombardment.

(b) Chemical methods

In spite of the expected difference in the reactivity of the primary and secondary ester groupings of triglyceride (III) no chemical reaction seems to have been evolved to distinguish between the two types of ester groups. Thus acid or alkaline hydrolysis, or reaction with a Grignard reagent[11] lead to mixture of products which point to hardly any selectivity. In one approach[8] the triglyceride (III) has been subjected to the acyloin condensation, and from the reaction mixture both simple and mixed acyloins (IV–X) have been isolated. The mixed acyloins (VII) and

R^2CH_2COO──┼──H   ⟶

$$\begin{array}{c} CH_2OCOCH_2R^1 \\ | \\ | \\ CH_2OCOCH_2R^3 \end{array}$$

(III)

$\Biggl\{ \begin{array}{lll} R^1CH_2CO\text{—}CH(OH)CH_2R^1, & R^2CH_2COCH(OH)R^2, & R^3CH_2COCH(OH)R^3 \\ \quad\quad\quad (IV) & (V) & (VI) \\ R^1CH_2COCH(OH)R^2, & R^1CH(OH)COCH_2R^2 & \\ \quad\quad (VII) & \quad\quad (VIII) & \\ R^2CH_2COCH(OH)R^3, & R^2CH(OH)COCH_2R^3 & \\ \quad\quad (IX) & \quad\quad (X) & \end{array}$

(VIII) indicate acyl groups in the *sn* 1- and 2-positions. Likewise acyloins (IX) and (X) indicate acyl groups in the *sn* 2- and 3-positions. The common acyl group, R^2CH_2CO, involved in the four mixed acyloin is the group at *sn* 2-position. Hence the other two acyl groups, R^1CH_2CO and R^3CH_2CO must be at *sn* 1- and 3-positions. This acyloin reaction of triglycerides is, however, not convenient to perform and monitor. It requires sodium–potassium alloy as the condensing agent, boiling xylene as solvent, and a relatively large amount of triglyceride.

(c) *Enzymic method*

Pancreatic lipase (obtained from hog pancreas) regiospecifically hydrolyses[9,10] ester groupings in the *sn* 1- and 3-positions, leading to, first a mixture of the *sn* 1,2- and *sn* 2,3-diglycerides (XI, XII), and eventually the *sn* 2-monoglyceride (XIII) and the corresponding fatty acids.

$$\begin{array}{c} CH_2OCOCH_2R^1 \\ R^2CH_2COO\!-\!\!\!\!-\!\!\!-H \\ CH_2OCOCH_2R^3 \\ (III) \end{array} \xrightarrow{\text{Pancreatic lipase}} \begin{array}{c} R^3CH_2CO_2H \\ + \\ R^1CH_2CO_2H \end{array}$$

$$\begin{array}{c} CH_2OCOCH_2R^1 \\ R^2CH_2COO\!-\!\!\!\!-\!\!\!-H \\ CH_2OH \\ (XI) \end{array} \quad + \quad \begin{array}{c} CH_2OH \\ R^2CH_2COO\!-\!\!\!\!-\!\!\!-H \\ CH_2OCOCH_2R^3 \\ (XII) \end{array}$$

$$\downarrow \text{Further hydrolysis Pancreatic lipase}$$

$$\begin{array}{c} CH_2OH \\ R^2CH_2COO\!-\!\!\!\!-\!\!\!-H \\ CH_2OH \\ (XIII) \end{array} \quad + \quad R^1CH_2CO_2H + R^3CH_2CO_2H$$

The enzymic method requires small amounts of the triglyceride, and the reaction is carried out very rapidly at an alkaline pH. Isolation of the free fatty acids reveals the identity of the occupants in the *sn* 1- and 3-positions. Isolation of the monoglyceride, and its subsequent hydrolysis, and identification of the resultant fatty acid show the acyl group in *sn* 2-position.

B. Methods for finding acyl groups attached at each of the three *sn* positions

The following methods illustrate how the joint applications of chemical and enzymic reactions help to forge very powerful tools to unravel the constitution of triglycerides and phospholipids. It was mentioned that the reaction of a triglyceride with a Grignard reagent leads to a mixture of products arising from both the primary and secondary ester groups. Therefore, the Grignard reaction, by itself, cannot help in the regiospecific distinction of acyl groups at *sn* 1,3- and *sn* 2-positions. The pancreatic lipase, as mentioned above, does enable this, but further direct distinction between the *sn* 1- and *sn* 3- acyl groups is not possible. (The pancreatic lipase method also cannot be applied to marine oils containing 20:5 or 22:6 acids.) However, if the initially produced mixture of *sn* 1,2- and *sn* 2,3-diglycerides from lipase hydrolysis is separated, and chemically transformed into the corresponding phospholipids, then with the help of other enzymes, complete stereospecific information can be obtained (see below).[12] Now enzymic reactions are fast, and it is not easy to arrest the lipase hydrolysis at the first stage to enable isolation of mixtures of diglycerides free from other by-products. It is found easy and practicable to isolate[11] by TLC not only the mixture of *sn* 1,2- and *sn* 2,3-diglycerides, but also the *sn* 1,3-diglyceride from the products of the Grignard reaction which thus proves useful after all.

(a) Method involving Grignard reaction, phosphorylation and phospholipase A

By this comprehensive method[11-14] information about the acyl groups at all the three *sn* positions is obtained (see Scheme 1). The triglyceride is reacted with either CH_3MgBr or C_2H_5MgBr, and from the reaction product the *sn* 1,2-/2,3- and *sn* 1,3-diglycerides are isolated by careful TLC (see details in appendix). The two groups of diglycerides are then separately converted into the corresponding phosphatides by reaction with phenyldichlorophosphate, $\underset{\underset{O}{\|}}{\text{PhOPCl}_2}$, in the presence of pyridine.

Now it has been known[15-18] that phospholipase A from snake venom shows stereospecificity in the partial deacylation of phosphatides.

SCHEME 1
BROCKERHOFFS' PROCEDURE[11-13] FOR DETERMINING ACYL GROUPS IN THE SN 1,2- AND 3-POSITIONS

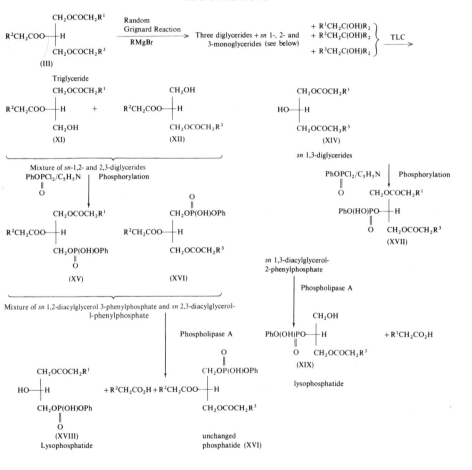

(i) *Determination of acyl groups at sn 2- and sn 1-positions.* Of the two phosphatides (XV) and (XVI), only the former (XV) is hydrolysed by phospholipase A, giving the free fatty acid (released from the *sn* 2-position), the lysophosphatide (XVIII) and the unchanged phosphatide (XVI). Separation of the products, and analysis of fatty acids as methyl esters show which acyl groups are present in the *sn* 2- and *sn* 1-positions, and by difference the acyl group at the *sn* 3-position is found.

(ii) *Determination of acyl groups in sn 1- and sn 3-positions.* The *sn* 1,3-diacylglycerol 2-phenylphosphate (XVII) is hydrolysed stereospecifically by phospholipase A, yielding the free fatty acid released from the *sn* 1-position and the lysophosphatide (XIX). Separation of the products, and analysis of the fatty acids as methyl esters show which acyl groups are present in the *sn* 1- and *sn* 3-positions. The acyl group in the *sn* 2-position in the parent triglyceride is found by difference.

(b) *Method involving Grignard reaction, phosphorylation and phospholipase C*

This method[19] is similar to the method (a) above, employing phospholipase A. As before the triglyceride is reacted with a Grignard reagent, and the mixture of *sn* 1,2- and *sn* 2,3-diglycerides is isolated from the reaction products by TLC, and converted into the corresponding phosphatidyl cholines (see Scheme 2) which are subjected to the action of phospholipase C (ex *Clostridium welchii*, type 1). It is observed that the *sn* 1,2-diacylglycerol-3-phosphatidylcholine (XX) is hydrolysed completely in 2 min, whereas the *sn* 2,3-diacylglycerol-1-phosphatidylcholine (XXI) requires over 4 h to do so—the final product in each case being the corresponding diacylglycerol. However, if the reaction is interrupted after 2 min, the products will comprise principally the *sn* 1,2-diacylglycerol (XI) and the unreacted *sn* 2,3-diacylglycerol-3-phosphatidyl choline (XXI) which are separated by TLC, and their fatty acids identified as methyl esters after transesterification. The fatty acid sequence in the original triglyceride can thus be determined. It can be further confirmed by GC–MS analysis of the t-BDMS-ethers of the *sn* 1,2- and *sn* 2,3-diacylglycerols (XI, XII).

(c) *Method involving pancreatic lipase (or Grignard reaction) and diglyceride kinase*

Pancreatic lipases help, as mentioned before, in finding out the acyl group in the *sn* 2-position, and also in preparing a mixture of *sn* 1,2- and *sn* 2,3-diacylglycerols. When such a mixture of the diglycerides (XI, XII) is subjected to the action of diglyceride kinase (ex. *Escherichia coli*) (in presence of ATP, mixed bile acid-salts, magnesium chloride and sodium

SCHEME 2
MYHER–KUKSIS' PROCEDURE[19] FOR DETERMINING ACYL GROUPS IN sn 1,2- AND 3- POSITIONS

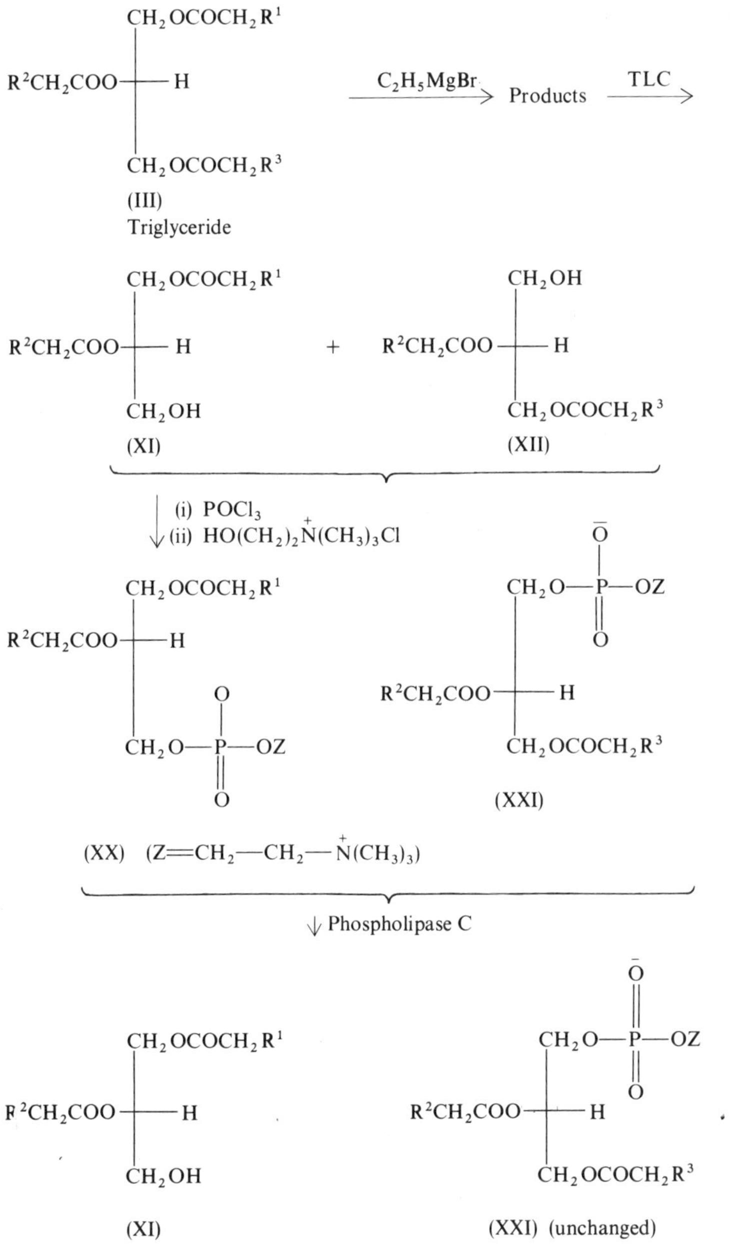

$$\begin{array}{c} CH_2OCOCH_2R^1 \\ R^2CH_2COO\!-\!\!\!-\!H \\ CH_2OCOCH_2R^3 \\ \text{(III)} \\ \text{Triglyceride} \end{array} \xrightarrow{C_2H_5MgBr} \text{Products} \xrightarrow{TLC}$$

$$\begin{array}{c} CH_2OCOCH_2R^1 \\ R^2CH_2COO\!-\!\!\!-\!H \\ CH_2OH \\ \text{(XI)} \end{array} + \begin{array}{c} CH_2OH \\ R^2CH_2COO\!-\!\!\!-\!H \\ CH_2OCOCH_2R^3 \\ \text{(XII)} \end{array}$$

$$\downarrow \begin{array}{l} \text{(i) } POCl_3 \\ \text{(ii) } HO(CH_2)_2\overset{+}{N}(CH_3)_3Cl \end{array}$$

$$\begin{array}{c} CH_2OCOCH_2R^1 \\ R^2CH_2COO\!-\!\!\!-\!H \\ \overset{O}{\underset{O}{\parallel}} \\ CH_2O\!-\!P\!-\!OZ \\ \text{(XX)} \end{array} \qquad \begin{array}{c} \overset{\bar{O}}{\underset{\parallel}{}} \\ CH_2O\!-\!P\!-\!OZ \\ \overset{\parallel}{O} \\ R^2CH_2COO\!-\!\!\!-\!H \\ CH_2OCOCH_2R^3 \\ \text{(XXI)} \end{array}$$

(XX) $(Z=CH_2\!-\!CH_2\!-\!\overset{+}{N}(CH_3)_3)$

↓ Phospholipase C

$$\begin{array}{c} CH_2OCOCH_2R^1 \\ R^2CH_2COO\!-\!\!\!-\!H \\ CH_2OH \\ \text{(XI)} \end{array} \qquad \begin{array}{c} \overset{\bar{O}}{\underset{\parallel}{}} \\ CH_2O\!-\!P\!-\!OZ \\ \overset{\parallel}{O} \\ R^2CH_2COO\!-\!\!\!-\!H \\ CH_2OCOCH_2R^3 \\ \text{(XXI) (unchanged)} \end{array}$$

phosphate buffer) selective phosphorylation of the *sn* 1,2-diacylglycerol (XI) takes place leaving the 2,3-diacylglycerol (XII) unaffected. After quenching the enzymic reaction, the *sn* 1,2-diacylglycerol-3-phosphatidic acid (XII) and the unchanged *sn* 2,3-diacylglycerol (XII) are isolated and separated by TLC, and the fatty acids in each of them are identified as methyl esters after transesterification. The fatty acids in the *sn* 1- and 3-positions are thus determined (see Scheme 3).

SCHEME 3
LANDS AND CO-WORKERS' PROCEDURE[20] FOR DETERMINING ACYL GROUPS IN THE *sn*-1,2 AND 3-POSITIONS

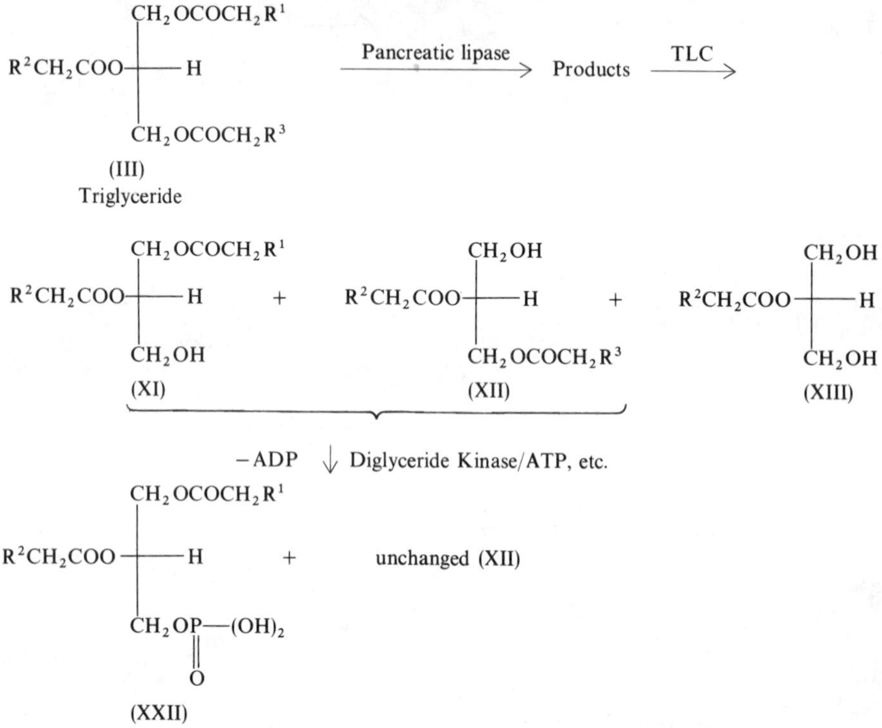

$$R^2CH_2COO-\overset{CH_2OCOCH_2R^1}{\underset{CH_2OCOCH_2R^3}{|}}-H \quad \xrightarrow{\text{Pancreatic lipase}} \text{Products} \xrightarrow{\text{TLC}}$$

(III)
Triglyceride

$$R^2CH_2COO-\overset{CH_2OCOCH_2R^1}{\underset{CH_2OH}{|}}-H \;+\; R^2CH_2COO-\overset{CH_2OH}{\underset{CH_2OCOCH_2R^3}{|}}-H \;+\; R^2CH_2COO-\overset{CH_2OH}{\underset{CH_2OH}{|}}-H$$

(XI)   (XII)   (XIII)

−ADP ↓ Diglyceride Kinase/ATP, etc.

$$R^2CH_2COO-\overset{CH_2OCOCH_2R^1}{\underset{CH_2OP-(OH)_2}{|}}-H \;+\; \text{unchanged (XII)}$$
$$\overset{\|}{O}$$

(XXII)

2.3 SELECTED APPLICATIONS

2.3.1 Positional Distribution of the Fatty Acids in the Triglycerides of some African Peanut varieties[21]

The analysis is performed on eight peanut varieties grown in Zaire and Senegal using pancreatic lipase, phosphorylation and phospholipase A (ex

Crotalus atrox). The fatty acids involved are 16:0, 18:0, 18:1, 18:2, 20:0, 20:1, 22:0, 24:0. Of these the first four constitute the major bulk. The saturated fatty acids and eicosenoic acid are found almost exclusively in the *sn* 1- and *sn* 3-positions—with palmitic acid being slightly more in the *sn* 1- than in *sn* 3-position. The 20:0, 22:0 and 24:0 acids are preferentially contained in the *sn* 3-position. Linoleic occurs preferentially in the *sn* 2-position, but oleic acid is distributed equally among the three positions. Among the various varieties of peanut oil from the same country, the variation in the percentage content of the above acid is not very large, but if the content of an acid in the peanut oil of one country is compared with the content of the same acid in the peanut oil of another country, then large differences are observed. This is particularly true of oleic acid—about 66% in the Bambey varieties of Senegal, but only about 43% in the Zairian varieties. Significant variations are found in the percentage content of the above acids in the *sn* 1- 2- and 3-positions of the triglycerides of the oil from the same country. There are recognisable differences in the percentage composition of the same fatty acid in the same *sn* position of the triglycerides of the oils from the two countries.

2.3.2 Stereospecific Analysis of Lard[19]

The total fatty acid composition found is similar to that reported[14] for pig outer back fat, using an old, and slightly different, method. The present authors[19] used method B(b) (see appendix). The acids making up the triglycerides of lard are 14:0, 16:0, 16:1, 18:0, 18:1, 18:2 and very small amounts (less than 1·5%) of 18:3 and 20:1. Palmitic acid is the major saturated acid which is confined largely at the *sn* 2-position. Oleic and linoleic acids are the major unsaturated acids and these are confined to the *sn* 1- and 3-positions, and are relatively missing from the *sn* 2-positions. Stearic acid, in contrast to palmitic acid, is present largely in *sn* 1-position.

2.3.3 Evaluation[22] of the Changes Induced by Heating Sunflower Oil at Various Temperatures in Air

Commercially refined Sunflower oil is treated in air (400–600 cm³/min) at 180° for 50, 70 and 100 h. Stereospecific analysis is performed on the unheated and heated oils, using pancreatic lipase. Significant differences in the fatty acid composition of the triglycerides of the two oils are observed. The heated oil, which contains products of thermal oxidation, is separated by column and thin layer chromatography over silica gel, into acylglycerol-monomers, -dimers, -trimers and -polymers. Each of these four fractions is subjected to stereospecific hydrolysis. It is found that after 70 h heating the amount of hydrolysis for the dimers is only half that of monomers; and that for the trimers being only a third—the polymers being the least hydrolysable by pancreatic lipase.

2.3.4 Correlation of Configuration of a Natural Triglyceride

Any optically active triglyceride can be synthesised[4] from D-mannitol; and therefore its configuration correlated via D-mannitol to D-glyceraldehyde. However, the final check of the identity of the synthetic and natural triglycerides is hampered by the extremely low optical rotations involved. Now enzymic methods enable us to determine the acyl groups at the three *sn* positions; and from this knowledge the configuration of a triglyceride can be deduced. However, such a result would appear to be based on the arbitrary standard of Hirschmann[5] (cf. plane project of glycerol with *sn* numbering). The enzymic method B(c)[21] is based on the experimental observation that diglyceride kinase phosphorylates one isomer of a pair of stereoisomeric unsymmetrical diglycerides. What evidence is there to assume that the isomer phosphorylated is *sn* 1,2-diacylglycerol? Lands and co-workers[20] have provided this evidence. By unambiguous synthesis they prepared the stereoisomers, *sn* 1,2-dipalmitoyl-, and *sn* 2,3-dipalmitoylglycerols, and subjected each isomer to the action of diglyceride kinase (ex *Escherichia coli*) in presence of ^{32}P-ATP (labelled in the γ-phosphate). Only the *sn* 1,2-dipalmitoyl glycerol led to the phosphatidic acid which incorporated the radioactivity. Therefore, the arbitrary assumption of *sn* 1,2-diacylglycerol is correct, and the determination of fatty acid sequence can tell us the configuration of a natural triglyceride.

2.4 APPENDIX: EXPERIMENTAL PROCEDURES

2.4.1 Method B(a) Involving Grignard Reaction Phosphorylation and Phospholipase A Grignard Reaction

To a vigorously stirred solution of corn oil (1·0 g) in dry ether (50 cm^3), is added, under anhydrous conditions, a 3 M ethereal solution of methyl magnesium bromide (2 cm^3). After 30 s of the addition, acetic acid (1 cm^3) is added; and after a further 30 s, water (10 cm^3) is added. The mixture is stirred for 2 min. The ethereal layer is separated, washed successively with water (10 cm^3), 2% aqueous sodium bicarbonate (10 cm^3), water

(10 cm^3), and dried over anhydrous sodium sulphate. The residue obtained on removal of ether is subjected to preparative TLC using 100 mg each time. The TLC plates (20 × 20 cm) are coated with a slurry, made from 25 g silica gel G and 40 cm^3 3% aqueous boric acid, dried in air and activated at 110° for 30 min. (Boric acid is used here, and later in the small scale Grignard reaction, to prevent acyl migration in the diglycerides.) The plates are developed first in ether–light petroleum (b.p. 40–60°) (8:92), and then in ether–light petroleum (40:60). In the developed plate, bands are spotted by dichlorofluorescein. The 1,3-diglyceride band appears between those of sn 1,2-/2,3-diglycerides and the tertiary alcohols (by-products). The glyceride bands are scraped out, extracted and rechromatographed using ether–light petroleum (1:1). The yield of sn 1,3-diglycerides obtained[13] is 60 mg. The same procedure applies for the isolation of sn 1,2-/2,3-diglycerides.

Small scale Grignard reaction which has been used for the isolation of sn *1,2-/2,3-diglycerides*[14]

To a solution of the triglyceride (40 mg) in dry ether (2 cm^3) is added, under anhydrous conditions, a freshly prepared 0·5 M solution of ethyl magnesium bromide (1 cm^3). After shaking the mixture for 1 min, glacial acetic acid (0·05 cm^3), and then water (2 cm^3) are added. The ethereal layer is separated, washed successively with dilute potassium carbonate and water, and dried over anhydrous magnesium sulphate. The product obtained on removal of ether is separated by preparative TLC on silica impregnated with 5% (w/w) boric acid (see comment above) using ether–hexane (1:1) as the solvent system—visualisation and further isolation of the sn 1,2-/2,3-diglycerides being done in a manner similar to the one mentioned above, yield, 6–7 mg (20–25%).

Phosphorylation[13]

A solution of sn 1,3-diglyceride (55 mg) in dry ether (1 cm^3) is added to a solution of phenyldichlorophosphate (0·25 cm^3), anhydrous pyridine (1 cm^3) and dry ether (1 cm^3). The reaction mixture is allowed to stand at room temperature for 90 min, then with cooling, several drops of water are added, followed by further addition of water (25 cm^3), methanol (30 cm^3), chloroform (30 cm^3) and triethylamine (1 cm^3). The resultant mixture is shaken thoroughly. The chloroform layer is removed, and the solvent distilled off at 40°, first at the water pump and finally under high vacuum. The residue, which contains only traces of the unchanged sn 1,3-diglyceride, is used as such in the next step. N.B. The phosphorylation of the sn 1,2-/2,3-diglycerides can be performed similarly. However, see method B(b) below.

Stereospecific hydrolysis[13] with phospholipase A

To the phospholipid, dissolved in ether, are added a solution (15 cm^3) of triethylamine saturated with carbon dioxide, 0·1 M aqueous solution of calcium chloride (1 drop), snake venom (*Crotalus atrox*) (1 mg); the mixture is shaken overnight under nitrogen. Isobutanol (20 cm^3) is added to prevent foaming, and the mixture evaporated below 40°, first at the water pump and then under high vacuum. The lipids from the residue are extracted into a mixture (1 cm^3) of chloroform–methanol (1:1) containing acetic acid (1 drop), and the extract applied (at the rate of 20 mg per plate) to a TLC plate (10 × 20 cm) coated with silica gel (10 g) and calcium sulphate (0·4 g). The plate is first developed in ether–light petroleum (35:65), dried in air for 5 min, kept over concentrated aqueous ammonia for 10 min and then developed in concentrated aqueous ammonia–methanol–ether (2:14:84). The lipid bands are visualised with dichlorofluoroscein, and eluted with chloroform–methanol (1:1). The approximate R_f values for the liberated fatty acid, the lysophosphatide and the unchanged phosphatide are 0·7, 0·05 and 0·4 respectively.

2.4.2 Method[19] B(b) Involving Grignard Reaction, Phosphorylation and Phospholipase C Grignard Reaction

The triglyceride (5–15 mg), dissolved in dry ether (0·4 cm^3) is reacted for 25 s with 1 M ethylmagnesium bromide (0·1 cm^3) then after adding acetic acid–ether (1:9) (0·4 cm^3), the mixture is shaken thoroughly for 30 s. Ether (4 cm^3) and water (0·5 cm^3) are then added, and the mixture shaken for further 2 min. The ethereal layer is separated, and washed successively with 2% aqueous sodium carbonate solution (0·5 cm^3) and water (2 × 0·5 cm^3). It is passed twice through a Pasteur pipette filled with anhydrous sodium sulphate, and the ether removed under nitrogen. The sn 1,3-diglyceride, and the sn 1,2-/2,3-diglycerides are isolated by preparative TLC on borate impregnated silica gel using chloroform–acetone (97:3) as solvent system.

Phosphorylation[19]

To the mixture of sn 1,2-/2,3-diglycerides (1–10 mg), contained in a 15 cm^3 tube provided with a Teflon-lined screw cap, is added a chilled solution (0·65 cm^3) of chloroform–pyridine–phosphorus oxychloride (47·5:47·5:5) (v/v). The resultant mixture is shaken well and allowed to stand at 0° for 1 h, and at 25° for another hour. The mixture is then poured into another similar tube containing dry powdered choline chloride (200 mg) and small bar magnet. The tube is closed tightly, and the mixture stirred vigorously at 30° for 15 h. Water (20 µl) is then added, and the mixture stirred further for 30 min. Chloroform and pyridine are evaporated as much as possible in a stream of nitrogen, the products extracted with a solution (12 cm^3) of chloroform–methanol–water–acetic acid (50:39:10:1) (v/v), and partitioned with 4 M ammonia. The aqueous phase is extracted once more with the above solution (4 cm^3). The combined extract is passed through a Pasteur pipette filled with anhydrous sodium sulphate, and then evaporated to dryness in a stream of nitrogen. The resultant phosphatidylcholines are isolated and purified by chromatography over silica gel H using chloroform–methanol–acetic acid–water (75:45:12:6) as the solvent system.

Stereospecific hydrolysis[19] with phospholipase C

A mixture of phosphatidyl cholines (0·5 ± 0·1 mg), ether (1 cm^3) and buffer solution (1 cm^3) (17·5 mM (trishydroxymethyl) ammonium methane solution (which is adjusted to pH 7·3) and which contains 1·0 mM of CaCl$_2$), is vortexed in a 15 cm^3 Teflon-lined screw capped tube for 15 s. Then 2 units of phospholipase C (ex *Clostridium welchii*, Type 1) dissolved in the buffer solution (2 cm^3) are added, and the mixture vortexed again for 15 s, and then shaken at 32° mechanically for 2 min. The mixture is chilled in an ice-bath, and the lipids are extracted three times with chloroform–methanol (2:1) (6 cm^3, 2 × 2 cm^3). The combined extracts are passed through a Pasteur pipette filled with anhydrous sodium sulphate, and evaporated to dryness in a current of nitrogen. The product is separated by preparative TLC on a plate (20 × 20 cm) coated with silica gel H. The plate is developed to a height of 9 cm with chloroform–methanol–acetic acid–water (75:45:12:6), and after drying for a brief period, it is developed to a height of 15 cm with heptane–isopropyl ether–acetic acid (60:40:4). The *sn* 1,2-diacylglycerol and the unchanged *sn* 2,3-diacylglycerol-3-phosphatidylcholine bands are scraped into two separate 15 cm^3 Teflon-lined screw capped tubes. Sulphuric acid (6% in methanol) is added to each tube which is closed tightly, and heated at 80° for 2 h in order to convert the lipid fatty acids into their methyl esters. This conversion can also be performed by reacting the lipids from TLC, with 1 M sodium methoxide (200 µl) at room temperature for 10 min. A known amount of the internal standard, methyl heptadecanoate in light petroleum is added to the reaction mixture which, after cooling in ice, is extracted with light petroleum (6 cm^3) and 4 M ammonia (15 cm^3), and the esters analysed.

Generation[19] of sn 2-monoglyceride by pancreatic lipase

The triglyceride (1 mg) is mixed with a buffer solution (0·5 cm^3) (prepared from 1 M-tris (hydroxymethyl) aminomethane, adjusted to pH 8·0, fortified with 10% gum arabic and 56 µl of a 45% aqueous calcium chloride). To it is then added porcine pancreatic lipase (2 mg) dissolved in the same buffer solution (2 cm^3). The mixture is vortexed for 30 s and then shaken at 37° for 10 min. The solution is then extracted with ether (3 × 2 cm^3). The extract is dried over anhydrous sodium sulphate, and then evaporated to dryness in a current of nitrogen. The monoglyceride is obtained by TLC on borate impregnated silica gel using chloroform–acetone (88:12) (v/v) as the developing solvent.

2.4.3 Method B(c) Involving Pancreatic Lipase (or Grignard Reaction) and Diglyceride Kinase[20]

As mentioned before, the starting material for this method is a mixture of *sn* 1,2- and 2,3-diglycerides which can be prepared by Grignard reaction or by the action of pancreatic lipase. The former procedure has already been mentioned.

Preparation of sn 1,2- and 2,3-diglycerides using pancreatic lipase

In a test tube (13 × 100 mm) are placed the triglyceride (2–5 mg), 1 M sodium chloride (0·15 cm^3), 1 M tris hydrochloride (pH 8·05) (0·10 cm^3) and lipase (0·05 cm^3, containing 2·7 units/cm^3, and isolated from steapsin). The reaction mixture is vortexed at room temperature for 3 min, then stopped by the addition of 1 N hydrochloric acid (0·20 cm^3). The lipids are extracted successively with chloroform–methanol (2:1) (v/v) (1·3 cm^3) and chloroform (0·80 cm^3). The combined extracts are evaporated, and the residue dissolved in chloroform–methanol (2:1) (0·10–0·20 cm^3) and put on the preparative TLC plate (coated to a thickness of 0·25 mm with reagent grade (200 mesh) silica gel, activated at 105° for 1 h. The plate is developed with 60% ether in light petroleum (b.p. 40–60°) to about 4 cm from origin, dried in air for at least 15 min, and developed again to the top, in 12% ether in light petroleum. The bands are visualised with 1% iodine in methanol (*and not with dichlorofluorescein which is carried with the diglycerides and inhibits kinase*). The diglyceride and monoglyceride bands are scraped into small columns and each eluted with 5% methanol in ether (10 cm^3).

Stereospecific phosphorylation with diglyceride kinase

The above diglyceride eluate is mixed with solutions of bile acids (10 µl of 200 mg/cm^3), ATP (0·10 cm^3 of 0·05 M), magnesium chloride (0·05 cm^3 of 1·0 M), sodium phosphate buffer (0·05 cm^3 of 0·50 M; pH 7·95), crude diglyceride kinase (0·10 cm^3 of 8 mg/cm^3). The mixture (pH 7·0) is incubated at 37° for 1 h with constant shaking. 1 M hydrochloric acid (0·20 cm^3) is then added, and the lipids are extracted successively with chloroform–methanol (2:1) (2·0 cm^3) and chloroform (1·3 cm^3). The combined extract, to which one drop of triethylamine has been added, is evaporated to dryness; and the residue, dissolved in chloroform–methanol (2:1) (0·1–0·2 cm^3) is applied on to the TLC plate (details as mentioned above) which is then developed to the top with 20% ether in light petroleum (b.p. 40–60°), allowed to dry for over 15 min, and then redeveloped to a height of 12 cm with chloroform–ethanol–formic acid (100:10:5) and the bands are visualised with dichlorofluorescein. The newly produced phosphatidic acid band is scraped off and eluted with 10% methanol in ethanol. To the residue obtained on removal of solvents, methyl pentadecanoate ($\sim$ 150 mµ moles) (internal standard) and 0·5 N sodium methoxide solution (2 cm^3) in methanol are added, and the mixture shaken well, and after 10 min the transesterification is stopped by adding 6 M hydrochloric acid (0·2 cm^3). Then light petroleum (5 cm^3) and water (5 cm^3) are added, and the mixture vortexed. The petroleum layer is dried over anhydrous Na_2SO_4–$NaHCO_3$(2:1) (w/w), and the resultant methyl esters analysed by GC using 10% ethylene glycol succinate (on Gas Chrom P) Column (8 ft × ¼ inch).

2.5 REFERENCES

1. BÖMER, A. and EBACH, K. (1928). *Z. Untersuch. Lebensm.*, **55**, 501.
2. MURATA, T. (1977). *Anal. Chem.*, **49**, 2203.
3. DEKI, M., KATO, T. and YOSHIMURA, M. (1975). *Kanzei Chuo Bunsekishoho*, **15**, 105; (1970). *Chem. Abst.*, **87**, 4030m.
4. BHATI, A., HAMILTON, R. J. and STEVEN, D. A. (1980). In *Fats and Oils: Chemistry and Technology*, Ed. R. J. Hamilton and A. Bhati, Elsevier Applied Science, London, 59 (see also original references cited therein).
5. HIRSCHMANN, H. (1960). *J. Biol. Chem.*, **235**, 2762. (See also IUPAC-IUB Commission on Biochemical Nomenclature (1967). *European J. Biochem.*, **2**, 127.
6. RHYAGE, R. and STENHAGEN, E. (1960). *J. Lipid Res.*, **1**, 382.
7. BARBER, M., MERRIEN, T. O. and KELLY, W. (1964). *Tetrahedron Letters*, 1063.

8. Paper presented by A. Bhati at the Symposium organised by the Lipid Group of Royal Society of Chemistry, Cambridge, 1982.
9. SAVARY, P., FLANCY, J. and DESNUELLE, P. (1957). *Biochim. Biophys. Acta*, **24**, 414.
10. MATTSON, F. H. and LUTTON, E. S. (1958). *J. Biol. Chem.*, **233**, 868.
11. YURKOWSKI, M. and BROCKERHOFF, H. (1966). *Biochim. Biophys. Acta*, **125**, 55.
12. BROCKERHOFF, H. (1965). *J. Lipid Res.*, **5**, 10.
13. BROCKERHOFF, H. (1967). *J. Lipid Res.*, **7**, 167.
14. CHRISTIE, W. W. and MOORE, J. H. (1969). *Biochim. Biophys. Acta*, **176**, 445; (1970). **210**, 46.
15. LONG, C. and PENNY, I. (1957). *Biochem. J.*, **65**, 382.
16. VAN DEENEN, L. L. M. and DE HAAS, G. H. (1963). *Biochim. Biophys. Acta*, **70**, 538.
17. TATTRIE, N. H. (1959). *J. Lipid Res.*, **1**, 60.
18. HANAHAN, D. J., BROCKERHOFF, H. and BARRON, E. J. (1960). *J. Biol. Chem.*, **235**, 1917.
19. MYHER, J. J. and KUKSIS, A. (1979). *Can. J. Biochem.*, **57**, 117.
20. LANDS, W. E. M., PIERINGER, R. A., SLAKEY, P. M. and ZSCHOCKE, A. (1966). *Lipids*, **1**, 444.
21. VAN PEE, W., VAN HEE, J., SONI, L. and HENDRIKX, A. (1979). *J. Am. Oil Chem. Soc.*, **56**, 901.
22. YOSHIDA, H. and ALEXANDER, J. C. (1983). *Lipids*, **18**, 611.

3

Industrial Aspects of Lipid Oxidation

J. C. ALLEN

Research Division, North East Wales Institute, Deeside, Clwyd, Wales, UK

3.1 INTRODUCTION

Despite the excellent efforts of Food Technologists, lipid oxidation remains a problem in many areas of food manufacture even today, and constant vigilance must be observed to ensure that its effect is minimised. The purpose of this paper is to attempt to relate the theoretical and laboratory findings on lipid oxidation to the industrial scene. It is neither a catalogue of disasters nor an exhaustive list of 'do's and don'ts', but endeavours to set down general points which are relatable to the theory of the subject.

The fundamentals of the problem can be stated quite simply:

1. The Arrhenius Activation Energy for autoxidation of polyunsaturated fatty acids is low—only 20 kJ/mol for the first step and around 40 kJ/mol for the second. The consequence of this is that the reaction rate is not significantly diminished by lowering the temperature of storage.

This is not to suggest that heating foods containing polyunsaturated fatty acids will not accelerate the oxidation of the fats therein, but it does imply that storage of such foods in freezers will have surprisingly little influence on the time needed before the onset of detectable oxidative rancidity, unless air is rigorously excluded. Freezer storage obviously, however, reduces bacterial degradation.

2. The rate of lipid oxidation is greatly accelerated by a number of catalysts acting as initiators in the process. These include transition metal ions such as Fe^{2+} and Cu^{2+}. When the metals are complexed, especially to nitrogen-containing ligands such as occur in metallopor-

phyrins such as haem, their catalytic efficiency greatly increases. The relevance of this to lipid oxidation in meat should be apparent. Perhaps surprisingly, denatured enzymes (such as peroxidase) can in some circumstances initiate lipid oxidation at a greater rate than they can in their native state. So paradoxically, heat treatment such as blanching can sometimes have an enhancing effect on oxidative spoilage. Also, light, in conjunction with a sensitiser such as chlorophyll or myoglobulin, can also initiate lipid oxidation. Methyl linoleate reacts with singlet oxygen at a rate more than 15 000-fold greater than it does with ground state oxygen. Also, specific enzymes such as lipoxygenases can catalyse hydroperoxide formation. These occur particularly in leguminous seeds, and is one of the enzymes which is inactivated by blanching.

There are thus, incidentally, three ways by which metalloproteins can promote lipid oxidation: through their native action, through the denatured molecule acting directly on a lipid reactant, and as a photosensitiser. All this may sound as if fat-containing foods can scarcely escape oxidative rancidity. This is manifestly not so. Even before the days

TABLE 3.1
LIPID OXIDATION FLAVOURS

	Flavour thresholds (ppm) in:		
	Water	Milk	Paraffin oil
Alcohols			
Ethanol	200		
1-Propanol	45		
1-Butanol	7·5	0·5	
1-Hexanol	2·5	0·5	
1-Heptanol	0·52		
1-Nonanol	0·086		
1-Decanol	0·18		
1-Penten-3-ol	3·0	3·0	4·2
1-Octen-3-ol	0·001	0·01	0·0075
Aldehydes			
Hexanal	0·03	0·05	0·6
Heptanal	0·03	0·12	0·04
Octanal	0·047	0·46	0·6
Nonanal	0·045	0·22	0·32
Decanal	0·007	0·24	1·00

Selected from Forss.

of modern technology, natural cooking acted as a protectant. In addition, many foods contain natural antioxidants such as plant phenols, which protect them against oxidative spoilage. Nevertheless, the facile nature of lipid oxidation makes it imperative for the food industry to take careful steps to avoid it. On the whole, a splendid record has been achieved, but a number of disasters have occurred, which are a salutary reminder to maintain standards.

3. Some of the secondary oxidation products of lipid oxidation, the volatile aldehydes and ketones, are organoleptically detectable at the part-per-billion level. Tables 3.1 and 3.2 give an indication of this, but the threshold level depends significantly on the manner in which the compound is presented, i.e. the type of food being assessed (Table 3.2). Figure 3.1 depicts a GLC trace of the headspace vapour from a packet of potato crisps, with a clear hexanal peak from oxidative breakdown of the oil used.

These latter comments remind us that the hydroperoxides are only transitory intermediates, and decompose into various carbonyl and other compounds. Although a properly conducted Peroxide Value determination is a good guide to fat quality, an oxidised oil can be reprocessed to give a deceptively low Peroxide Value. In such cases, the presence of secondary products plus the lower level of antioxidant present, will enable further rapid oxidation to occur. For this reason it is preferable not to rely solely upon Peroxide Value as an index of oil quality, but undertake an Anisidine test in addition. Such a 'Totox' determination is very useful in the detection of reprocessed oils.

TABLE 3.2
CHARACTERS AND THRESHOLD LEVELS OF SELECTED ALIPHATIC ALDEHYDES

Aldehyde	Threshold (ppm) in paraffin oil	Character
cis-3-Hexenal	0·09	'Green bean'
trans-2-Hexenal	0·60	'Green'
cis-4-Heptenal	0·000 5	Cream to tallow
trans-6-Nonenal	0·000 35	'Hydrogenation'
trans-2-cis-6-Nonadienal	0·001 5	Beany
trans-2-trans-6-Nonadienal	0·02	Cucumber
trans-2-trans-4-Decadienal	0·10	Stale frying oil

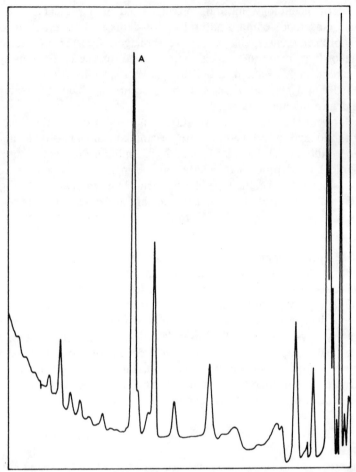

FIG. 3.1. Chromatogram of headspace vapour from potato crisps, showing presence of hexanal (peak A), a product of oxidative breakdown of fats. Column: 10% Silicone OV-101 on Diatomite CLQ Support. Carrier: N_2 at 40 ml/min. Column temperature programmed 20–180°C at increase of 4°C/min.

3.2 THE RAW MATERIALS

The UK alone imports several hundred thousand tonnes of edible oils a year, principally palm, soybean, rapeseed, sunflower, palm kernel, coconut and peanut oils (Table 3.3). As the world's economy develops, there is a trend amongst the primary producing countries to refine the oil

TABLE 3.3
IMPORT TO UK FOR HUMAN CONSUMPTION IN 1982
(tonnes)

Oil	
Palm	162 569
Soybean	123 934
Rapeseed	61 483
Sunflower	50 297
Palm kernel oil	28 626
Coconut	27 696
Peanut	10 432

themselves. If properly done, and if the oil is then transported under closely regulated conditions, it can be advantageous to the final quality of the oil. However, it is not always so easy to ensure adherence to specifications in such circumstances. For instance, there have been examples of the occurrence of lipid oxidation in transported oils, which have been tracked down to it having been pumped from dockside containers to ships via pipes with brass couplings or valves. Traces of copper from the brass were sufficient to initiate lipid oxidation in the oil.

Whilst considering crude oils, perhaps a point to bear in mind is the impact of new methods of crop growing and selection on the development of new oils. Low-erucic acid rapeseed oil is a good example. Its properties, including its oxidative stability, have been improved to the point where it can be successfully used as a component of frying oils in large-scale manufacture, which may even involve high moisture, and circulation at a temperature of up to 170°C, such as is used in potato crisp manufacture. Indeed, rapeseed oil is now sold as a high-quality retail product, with an 'up-market' image and price to match.

The bulk storage of oils and fats presents problems. The Oil Companies are generally very aware of these problems, and will readily offer excellent advice. However, disasters still occur. Heating the oil above the recommended temperature in an effort to improve its 'pumpability', is often the cause. For instance, the overheating by only 10°C of a blend containing sunflower oil was enough to cause severe rancidity in several tonnes of a spray-dried babymilk formulation. Although its rancid taste was obvious to any adult, a number of small infants were not so discriminating (or were not given any option), and cases of sickness were reported. The incident was rapidly dealt with before any serious illness occurred, but it serves as a reminder of the importance of adhering to manufacturing specifications and procedures.

There is still much room for improvement. The author is aware of factories currently storing susceptible oils at 60–65°C to increase their mobility. Poor liaison between plant operators and technical management can be to blame. One recent example is of a revised operating procedure being produced with all temperatures in degrees C, whereas the fairly ancient plant had all gauges calibrated in degrees F!

A good general rule is to check the Peroxide Value on all deliveries, and also on stock after long periods of shut-down, e.g. after long holidays. Also, the relatively minor amounts of oils which may be introduced into a blend must also be carefully checked. It should not be forgotten that 10 tonnes of good oil plus as little as 50 kg of rancid or partially-rancid oil can produce 200 tonnes of rejected product, making the assumption that the oil is incorporated into the product at a 5% level.

The use of butter is worth a mention. Butter can be very prone to both lipolytic and oxidative rancidity. 'Intervention' butter although commercially advantageous due to its relative cheapness, can be very old. It is produced under well controlled conditions, however. Intervention butter oil is more vulnerable to lipid oxidation, since the processing renders it free of antioxidant components.

The unpalatability of rancid oils fortunately means that there is little chance of large amounts being consumed. However, the reprocessing of heavily oxidised oils to remove peroxides and flavour volatiles can still leave behind significant amounts of potentially toxic residue, such as unsaturated keto-acids. The 'Spanish oil scandal' of 1981 was a frightening example of the tragic consequences of improper treatment of oil. A mixture of highly oxidised oils, principally rapeseed, olive seed and grape seed, had been adulterated with aniline and thereby designated for industrial use. However, it was somehow acquired by unscrupulous people and sold as an edible oil. Over 2000 people were affected in an epidemic of a disease resembling severe pneumonitis, and more than 300 died. The best evidence so far is that the disease was caused by a combination of the oxidation products plus anilides. Although there was some dispute about this, the evidence of pentane in the breath of patients, and malonaldehyde-positive reactivity of certain post-mortem samples, is good evidence for the involvement of toxic oxidation products in the disease. An issue arising from this horrifying incident is the need for the reputable blender of oils and fats to protect himself and the consumer from marketing and buying such oils.

3.3 THE FINISHED PRODUCT

So far, the emphasis has been on the raw material, the oils and fats. It is now appropriate to turn attention to the finished product, the food itself. There is a natural tendency to regard those foods with the highest lipid concentration as the most susceptible to lipid oxidation, and the most troublesome in this regard. This is not necessarily so.

Rancidity depends not only on the concentration of the lipid, but also on its reactivity and its environment. For instance, the lipid content of 'instant' potato powders is only around 1%, but they include a fair proportion of polyunsaturated fatty acids and are particularly susceptible to oxidation by their being dispersed on a powder with a large surface area. This increases the exposure of the lipid to oxygen. Such products have a very short shelf-life unless specific precautions are taken: this often includes adding extra ascorbic acid, which can also be used as a marketing aid ('with added Vitamin C').

It is difficult to summarise guidelines, apart from the obvious, for the avoidance of lipid oxidation in foods. Each product has its own special susceptibilities, depending on its composition, presentation, packaging and the processing conditions used. Cereal-containing products, for instance, can be spoilt by poor grain storage conditions, and in particular by being kept under conditions of high moisture. This can lead to fungal contamination, which can, through the action of fungal lipases, increase the amount of free fatty acids, which oxidise more rapidly than triglycerides. Incidentally, baking offers one rare positive effect of lipid oxidation: wheat flour is low in lipoxygenase activity, and legume flours like soybean or horsebean flour are added in order to bleach it. The action occurs via co-oxidation of carotene.

It is interesting that liquid milk is generally regarded as stable towards lipid oxidation, except when mistreated by, for instance, the addition of Cu^{2+}. However, work from the author's laboratory has indicated that the principal characteristic of milk described as 'stale' may well be an oxidised off-flavour. Normal processing of milk does not remove the factors which render it liable to lipid oxidation. On the contrary, separating, churning, drying and addition of other materials can often increase it, by causing the intermixing of hitherto separated catalysts, and by contamination from the processing plant. For instance, migration of Cu^{2+} into cream on churning can cause rapid flavour impairment. Buttermilk is often highly susceptible to lipid oxidation: it contains phospholipids, with a high proportion of polyunsaturated acyl groups, and certain of these (principally phosphatidylethanolamine) can bind metal ions in a pro-oxidative fashion, by forming a metal–phospholipid complex at the oil–water interface of the emulsion.

Butter itself can oxidise, both by metal-ion catalysis, which gives rise to 'fishy' taints, and by photo-oxidation. The latter often occurs when non-foil-wrapped butter is held too near to fluorescent lights, in display chill cabinets. Dried whole milk is stable provided it is kept dry. However, 'humanised' babymilks, formulated from milk proteins and vegetable fat blended to mimic the fat concentrations and polyunsaturated fatty acid levels of human milk, can give rise to serious problems. One of these arises from legislative demands to add Cu^{2+} for nutritional reasons (which in themselves seem rather dubious). Careful control has to be exercised over the order of additions, the temperature of mixing, the duration of blending and drying, and the type of packaging. It is imprudent and illegal to add antioxidants to such milks, and the only sure way to protect such a susceptible product is to gas-pack under nitrogen. Lack of care or awareness over packaging and presentation has also caused problems related to lipid oxidation: photo-oxidation of butter in display cabinets was mentioned above. Fatty products, such as potato crisps and snack foods, can also suffer from such problems, and there often has to be a compromise on print design between protecting the product from light, and displaying it attractively to the customer. Problems can occur in the most unexpected ways. For instance, a problem on lipid oxidation involving a well-known semi-sweet biscuit was eventually traced to a logo on the wrapping paper, which was printed in a gold colour. The oxidation and taint which resulted was eventually traced to the pigment in the ink, which consisted of a bronze powder with a 90% copper content. This, plus grease penetration into the wrapping, was enough to cause the damage by initiation of lipid oxidation.

Finally, one other aspect perhaps deserves mention. The current nutritional lobby for an increasing proportion of our fat intake to consist of unsaturated lipids could in principle rebound. We are absorbing a relatively increased amount of polyunsaturated fatty acids, despite calls for an overall reduction in total amounts of fat ingested. Thus, incidents of ingestion of lipid oxidation products may well occur. We are not aware of the effects of ingestion of very low levels of lipid oxidation products over long periods, much less the effect of acute ingestion of relatively large quantities of such compounds. However, mutagenic effects of a number of materials related to lipid hydroperoxides and secondary oxidation products have been reported. It may be that the pendulum has swung too far. Careful research is now needed in this area.

The Technical and Production Managers of UK food companies will not *openly* admit to having any problems with lipid oxidation whatsoever, and indeed, their general standard of excellence in this area is obvious. However, *privately*, many of them voice concern and are quick to raise practical examples. It is thus important that vigilance be always maintained.

4

Hydrogenation of Oils and Fats

H. B. W. PATTERSON
Consultant, Bebington, Merseyside, UK

4.1 INTRODUCTION

To obtain a sense of perspective in viewing current hydrogenation practice some indication of earlier techniques is worthwhile. The direction and rate of development is then more easily assessed; a more realistic speculation as to what future developments are economically feasible is possible. From a practical standpoint it is convenient to consider three areas of the subject.

(i) The environment we create for our work, that is the layout of the plant and services.
(ii) The way in which the three phases involved in the hydrogenation reaction, hydrogen gas, liquid oil and solid catalyst are brought together in bulk, that is the design of the autoclave or hydrogenation vessel.
(iii) The behaviour of the molecules when they have reached the position of being able to react with one another at the catalyst surface and what these several reactions may be.

4.2 LAYOUT

Since between 74·2% and 4·1% v/v hydrogen in air forms an explosive or at the very least a flammable mixture the importance of keeping outside of these limits has long been recognised. This means if hydrogen leaks out from any vessel ventilation must be such that it is rapidly dispersed and diluted. Being the lightest substance this is not difficult to arrange in plant or laboratory. If air leaks into a vessel it must not find sufficient hydrogen with which to ignite and in general sources of ignition must be kept away from locations where mixtures might occur, albeit accidentally. Hence electric motors are flame proofed, lights pressurised and when engineering work is to be done that could cause ignition the vicinity is first cleared of hydrogen. It is to be noted that an active nickel catalyst in a non-pyrophoric non-fatty powder form can promote ignition of hydrogen in air. Further, discarded, almost exhausted fatty catalyst, if left lying in a draught may commence to smoulder and eventually to inflame because the fat is encouraged to oxidise. It is a useful and common practice to have provision for vacuum on hydrogenation autoclaves. This is best provided by a steam ejector which is also probably the cheapest method. Such ejectors like safety pressure vents must discharge outside, not pointing at a nearby building. As a precaution against leaky valves on hydrogen or vacuum lines at the autoclave it is feasible to fit two block valves in series with a small vent cock between them. An inert gas, probably nitrogen, is most useful as a purge of the autoclave after withdrawing air by vacuum and before admitting hydrogen, or after withdrawing hydrogen by vacuum and before admitting air. The inert gas itself of course is withdrawn by vacuum before proceeding to the next operation. The use of steam in place of nitrogen has been known in a few plants but is much less popular than vacuum or the combination of vacuum and nitrogen just described. Immediately upstream of the hydrogen inlet control valve on the autoclave a non-return valve is sometimes fitted to act on those occasions when the hydrogen supply pressure falls temporarily below the pressure at which the autoclave has been working. In such a situation partly hardened oil would be driven some distance back into the hydrogen supply line and block it. Such a non-return valve must be biased in such a way that it closes when the excess of hydrogen supply pressure over autoclave working pressure becomes less than say, 0·5 atmospheres. It is not sufficient for the non-return valve to attempt to close at the moment reverse flow is commencing. Various flow control and pressure relief valves have important parts to play in the safe running of the plant. Obviously if the pressure on the suction line feeding a hydrogen compressor falls below a minimum 0·5 in water gauge the compressor must be switched off automatically. Failure to do this implies air is then being compressed into the hydrogen high pressure store. Table 4.1 shows a typical pressure control situation. In this table the central reducing valve is bringing down the pressure in the HP hydrogen store to whatever has been decided as the maximum

TABLE 4.1
EXAMPLE OF RELATIVE PRESSURE LEVELS ASSOCIATED WITH A HYDROGENATION AUTOCLAVE

Type of limit	Pressure (atm)	Control
Hydraulic test	9–12	
	8·3	Bursting discs
'Design pressure'	8	Individual safety valves
	6·6	Optional central relief valve
Maximum operating pressure	6	Central reducing valve
Usual operating pressure range	up to 5	

All autoclaves tested to withstand full vacuum.

operating pressure of the autoclaves (6 atms) whilst manual or instrumental control on each individual autoclave sets the operating pressure for any particular hardening batch. The optional central relief valve, if installed, would be located somewhere downstream of the central reducing valve so that if the latter failed to reduce adequately this valve would come into action before the invididual safety valves on autoclaves were obliged to vent pressure.[1]

A feature most obviously gaining in popularity in the last ten years, but not at all novel, is some means of energy saving by heat exchange since hydrogenation is an exothermic reaction liberating enough heat in dropping the iodine value (IV) one unit to raise the oil temperature by slightly more than 1·5°C. There appears to be slight variation in specific heat from one oil to another and in the same oil at different temperature levels. Some systems of saving energy have correctly been developed to save cycle time and hence substantially improve utilisation or productivity. The most elementary form of energy saving is to lead the cooling water which controls reaction temperature in the autoclave and later lowers it to below 100°C prior to filtration, to transfer its heat for some useful purpose such as warming a stock of crude oil waiting to enter the refinery for neutralisation, after which it reverts to its task of cooling further autoclaves. Such tasks save heat but not time in the hardening plant. If we can charge the autoclave with soft oil already at reaction temperature or nearly so, then at the end of hardening, drop the uncooled oil with its catalyst to an air free tank, we have saved heating time, cooling time and filtration time retaining perhaps some 15 min emptying time. Catalyst can be added during filling time. The saving on a 10 ton charge could be as given in Table 4.2.

TABLE 4.2
ENERGY/TIME SAVING MODIFICATION

	Before (h)	After (h)
Filling	0·5	0·5
Heating	1·5	—
Catalyst addition	0·25	—
Hydrogenation	2·00	2·00
Cooling (to 100°C)	0·75	—
Filtering/emptying	1·50	0·15
Total	6·50	2·75

Before money is spent on items of plant to double productivity or better, it must be established that the market exists for the additional production, that the extra soft oil and hydrogen are available to sustain it and that the capital expenditure will certainly be recovered in an acceptably short time. Numerous examples of failure to satisfy these criteria at the planning stage exist. Such enhancing of facilities allows the autoclave to concentrate on its primary design task of bringing gas, oil and catalyst into contact in the chosen conditions of temperature and pressure. The degree of agitation is not usually variable without resort to engineering modification. As suggested for the example in Table 4.2 a simple arrangement for a plant with possibly only one autoclave is to pass the soft oil through a closed coil in the drop tank containing the hot hardened oil and catalyst and thence into the waiting autoclave. Filtration of the hardened oil may then proceed at leisure, some further cooling to below 100°C being performed en route to the filter by means of a water cooled heat exchanger. The heat of reaction of the hydrogenation process may serve to raise the temperature of the oil to the final operating temperature at which it is controlled. Such a rise typically would be from 140°C to 180 ± 5°C. How much of this heat is available depends on how large is the IV drop. In the case of many fish oils the IV drop may often be as large as 110 units, so here is a substantial amount of heat to be transferred using the cooling system of the autoclave itself. An arrangement has been suggested wherein two autoclaves work in a roughly reciprocal fashion, as one empties the other fills, hardened and soft oils flowing on either side of a heat exchanger. Here however the duration of the respective filtering and filling times must be about equal and ideally should coincide. To escape from the constraints which this applies a further elaboration has been to provide a number of holding tanks of insulated nitrogen blanketed soft oil to supply the autoclaves of a larger plant whilst each autoclave has access to a drop tank or heat exchanger. In this way each hot charge which is dropped can be cooled to a very useful extent towards its filtering temperature of 100°C max, whilst time is saved and the incoming oil, as it is warmed, is accumulated for the next available autoclave. This arrangement caters for hardening cycles of appreciably differing lengths because of its greater flexibility. Production planning will still be needed to ensure its most efficient use. A number of possibilities are put forward by different plant suppliers.[2-9] No doubt suppliers are prepared to be flexible to meet the circumstances of a particular customer. Any proposed energy and time saving facilities must be closely geared to the complexity and possible variability of a production programme. This can range from one autoclave producing four different products during the week from two vegetable oils to 12 autoclaves working on lauric and other vegetable oils, fish oils and animal fats to produce 30 products in the week. Features which must be common to the simplest and most elaborate arrangements are:

(i) Structural design features must not encourage catalyst to deposit at bends, valves etc.
(ii) Cleaning must be easy.
(iii) Hold up of oil in plant components must not promote contamination of one product by another so as to put product specification at risk.

The layout of hardening plants has been influenced, particularly since the 1950s, by steadily improving facilities for the production of hydrogen at 99·8% purity or better even on the modest scale of ca. 50 m^3/h. Electrolytically generated hydrogen has been of a high standard of purity from the very early days of the fat hydrogenation industry,[1] so that the so-called 'dead end' technique of simply applying hydrogen pressure to the oil and catalyst in the autoclave and internal agitation sufficed. As hydrogen was absorbed more could enter. Electrolytic generators were always both flexible in themselves and capable of being produced as small units. Hydrogen prepared by other methods being less pure was circulated through the autoclave, scrubbed and returned. As in all systems an accumulation of inert impurities, such as nitrogen or methane, had to be purged to atmosphere with the loss of some hydrogen at the same time. Since the 1950s catalytic reaction of hydrocarbons with steam—hydrocarbon reforming—has reached the stage of providing very pure hydrogen at competitive cost even from the smallest units. Now that the

TABLE 4.3
HYDROGEN PURITY (ON DRY GAS CALCULATION)

Inert gases (N_2, CH_4)	0·55% v/v max. Commonly this is ca. 0·1% v/v in practice
Carbon monoxide	0·05% v/v max. Nickel carbonyl (Ni (CO_4)) becomes unstable at 110°C and breaks up completely at 160°C, by which temperature CO content is much less important. Often CO is under 10 ppm in practice
Sulphur	0 – 1 ppm S commonly attainable; more could be tolerated[1]
Mercury	100 µg/m^3 max. Much less is easily attainable[1]
Oxygen	Up to 150 ppm O_2 may be found in electrolytic hydrogen
Halogens	Nil

level of inert impurities is under 0·2% v/v the gas space above the oil in the crown of the autoclave can be made large enough (say 30% of total internal volume) to accommodate the build-up there of inert impurities; but even if they reach the 20% v/v level where the rate of hydrogenation begins to be impeded, a controlled purge to atmosphere for 1 – 2 min should bring about acceleration of the hardening and stimulate an increase in temperature if this had begun to fall. Table 4.3 shows the acceptable maxima for most of the common impurities in hydrogen which is to be used for any class of fat or fatty acid hydrogenation. The 'dead end' technique has been the dominant one for years. Compression of moist hydrogen (1·5% v/v H_2O say) to 7 atm followed by cooling brings down moisture content to ca. 0·1% v/v.

A convenient capacity for hydrogen storage is equivalent to 4 h maximum usage. Substantially longer interruptions can be met by building up stocks of hardened oil. Since pressure vessels have to be given a hydraulic test their foundations must be strong enough to bear their weight when filled with water.

There are four main classes of oils commonly processed in hydrogenation plants although most plants are not concerned with all four. These are lauric oils, other vegetable oils, marine oils and animal fats. Pipelines, filter units and other ancillaries are often the cause of one class contaminating the other unless steps are taken to enforce segregation which can easily entail, for example, separate pipelines, filters and autoclaves for lauric oils and marine oils. Centre points at different stages during the handling system from crude oil intake to distributing the various hardened oils to their store tanks make a notable contribution to guarding against human error, especially if at any particular junction only pipes handling the same class of oil are compatible with one another when making a union. Unions are available such that when brought together the pipeline becomes open and when the union is uncoupled the line is thereby closed. This allocation of lines and plant units to a particular class achieves a static segregation for a capital cost. Where segregation as an important safeguard for specification arises only at intervals the most economic way of achieving it is to practise a dynamic segregation. Having carefully cleared the way in advance, the special product is moved from stage to stage diverting the first few kilograms of flow whenever some contact with another oil may have been possible. An obvious time to plan such a manoeuvre is the weekly start-up of a clean plant. Standards continue to rise and analytical techniques become more searching. Where a genuine need to segregate is present, failure to spend effort and money to meet it leads to customers becoming distrustful and irritated that their specification, first accepted as feasible, is not met by the product.

4.3 AUTOCLAVE DESIGN

It is the hydrogen physically in solution in an oil which reacts chemically with it when both contact one another at the surface of the catalyst. Since increase of the oil/gas interface directly increases the opportunity of more hydrogen to dissolve and replace that which has been removed by combination, all manner of ways have been used to increase this interface. To the early system of bubbling a stream of gas up through the oil and returning that which escaped to the sparger below, stirrers of various shapes were added. Droplets of oil/catalyst were sprayed from the crown of one vessel to collect at the foot and be pumped to the crown of another and so on, so as to form in effect a crude continuous system in four or five vessels. An effective old design is one by Wilbuschewitsch in which hydrogen in the top gas space is continuously drawn into the throat of a centrally mounted suction tube by the injection into it of a circulating stream of oil as indicated in Fig. 4.1. Dispersal of hydrogen bubbles into oil rather than oil droplets into hydrogen has come to be universally preferred.

From the 1950s the turbine stirrer became very popular, especially in

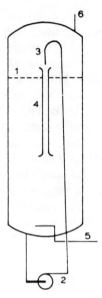

FIG. 4.1 Wilbuschewitsch Mixing Jet. 1, Normal oil level; 2, oil catalyst circulating pump, gland sealed with oil compatible with charge; 3, ejector; 4, oil/hydrogen mixing tube; 5, Bottom gas connection to sparger; 6, top gas (balance gas).

the USA, as a means of dispersing hydrogen into oil and for comparable applications in the chemical industry generally.[10] Figure 4.2 shows a typical radial flow flat blade turbine stirrer. On the circular disc of several inches diameter (according to the diameter of the autoclave) there are mounted a number of flat vertical discs as shown.

When the turbine is immersed and rotated (150 rpm is typical for some applications) streams of oil are expelled rapidly from the centre of the disc at both upper and lower surfaces. If gas bubbles are then admitted just below the disc these are drawn into the oil stream where there is intense shearing action of one layer of oil moving over another. The bubbles are shattered and dispersed. Oil streams strike the wall of the autoclave where a few vertical baffles are mounted to prevent swirling. Some of the oil is drawn back to the centre of the disc—above and below—to replace that just expelled. In a small autoclave (5 tonnes charge) one radial flow turbine would suffice. For larger autoclaves (ca. 10 tonnes) two would be mounted on the central shaft, the upper one creating some surface agitation. For a still larger vessel (15 tonnes) a third turbine is added a few inches below the working oil surface, but

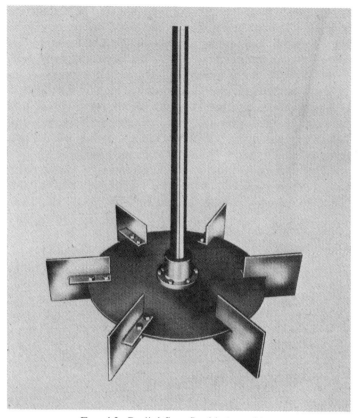

FIG. 4.2. Radial flow flat-blade turbine.

now the blades are inclined to the vertical to give a downward axial flow. This discourages escape of gas bubbles into the head space but also creates sufficient vortex effect to draw back hydrogen which may have done this. Most of the energy is consumed by the lowest turbine and least by the uppermost axial flow one which is smaller. As the oil temperature rises and dispersing bubbles permeate the charge, viscosity and overall density drop, hence the installed power on the turbine motor of 2 kW/tonne is not drawn but falls to a little over half that.[1,10] In the engineering laboratory it is possible to demonstrate the effect of the power used on the turbine and of the wall baffles. First with a stationary turbine power is consumed by forcing gas through the sparger and against the pressure in the autoclave. The flow pattern is of a rising stream of bubbles from the holes of the sparger. The turbine is set in motion and the applied power steadily increased. Up to the point where the power drawn by the turbine is about equal to that used to force the gas through the sparger the flow pattern is comparable, with numerous small bubbles breaking the surface. In fact it is about this point that the ratio of gas absorbed per unit of total power employed is highest. The rate at which gas is absorbed per unit volume of oil is far from the maximum and as the power applied to the turbine is increased several times the rate of gas absorption climbs rapidly. This is what is required; we have now bought time with energy and much shorter cycle times are obtained. It will be evident that as well as an intensive mixing of gas and oil the turbine continuously mixes the oil charge itself so that the hydrogenation effect is evenly distributed. Also since the 1950s the

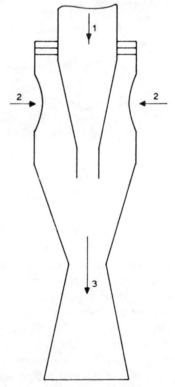

FIG. 4.3. Modern venturi mixing jet.[1] 1, Oil-catalyst stream; 2, hydrogen entry ports; 3, zone of intense shear.

venturi jet mixer has been developed for gas/liquid contacting and is now used in various chemical plants as well as in the hydrogenation of oils and fatty acids.[1,9] Figure 4.3 shows how hydrogen is drawn into the jet via lateral ports by the rapid downward flow of oil/catalyst in the central core. Before escaping into the body of the autoclave gas bubbles and oil experience intense shearing effects in the throat of the jet. Earlier, less sophisticated designs injected hydrogen from downward facing nozzles in the centre of the oil stream as it entered the venturi, but this feature has been abandoned in favour of the more effective mixing of the present design.

Vacuum is available as indicated in Fig. 4.4 and as in other systems can be applied during filling when it helps dry the charge. Cooling/heating is via an external heat exchanger. Accurate control of hydrogen fed to the autoclave is a feature offered by the supplier and this can be used to take into account any hydrogen absorbed during the final cooling by circulation. Figure 4.4 shows the arrangement of a one loop hydrogenation reactor.

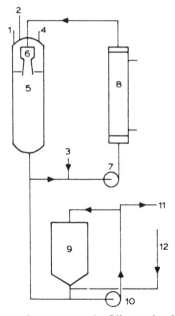

FIG. 4.4. Loop hydrogenation reactor. 1, Oil supply; 2, hydrogen supply; 3, catalyst suspension; 4, vacuum; 5, autoclave; 6, venturi jet; 7, circulating pump; 8, heat exchanger; 9, intermediate tank; 10, filter pump; 11, filter; 12, from first runnings tank.

Four to six reactors may be combined in series to give a continuous flow. Separate pumps transfer oil from one chamber to the next; in effect there is a common gas space. Continuous hydrogenation systems have become popular in general only for hydrogenation of fatty acids. This is probably partly because in that application long production runs giving the same product from the same or similar feedstock are usual.

When the concentration of adsorbed hydrogen on the nickel catalyst is high the chances of selective hydrogenation are lowered, that is to say some proportion of the unsaturated fatty acids in the triglycerides may be fully hydrogenated and some almost so, before markedly unsaturated fatty acids have had their due share of hydrogen. Both higher operating pressure and intense agitation increase passage of hydrogen to the nickel surface. However turbine and venturi jet have been accepted for selective hardening tasks. If a *trans* promoting effect is needed this can be obtained by the use of precisely prepared sulphur poisoned catalyst now commercially available. Suppliers of the loop hydrogenation reactor claim one of its outstanding advantages is its high catalyst economy arising from the intense agitation. For those who see an advantage the headspace of the loop reactor—and indeed the headspace of a turbine stirred autoclave—can be provided with a facility for abstracting a steady stream of hydrogen, drying it and returning it to the reactor.

4.4 REACTIONS AT THE CATALYST SURFACE

Crystallites of nickel metal produced in the reductive destruction of its compounds fall within the 40–100 Å size range. Whilst this gives an excellent surface area of $100 \text{ m}^2/\text{g}$ or more it presents an impossible burden for filtration. It is arranged therefore in the most favoured method of manufacture that the crystallites are generated within the pores of a siliceous material which supports them, hence the particle size of the product is now much greater. A good proportion of the particles fall in the 10 µm (100 000 Å) range and above, so filtration is no longer a problem. At the same time material and procedures are chosen to ensure that pore size is well above 25 Å diameter so as to allow passage in and out of triglyceride molecules with a possible dimension of 15 Å. Relatively wide and not very deep pores allow the easy escape of a triglyceride molecule after only one of its double bonds has been

---

1 Angstrom $(\text{Å}) = 10^{-10}$ m; 1 Nanometre $(\text{nm}) = 10^{-9}$ m; 1 micron $(\text{µm}) = 10^{-6}$ m.

hydrogenated. In a deep pore the chances of more than one double bond being hydrogenated during one visit to the catalyst are much greater. The first situation favours a controllable progressive partial hydrogenation of the more unsaturated fatty acid groups which in the earlier parts of the reaction dominate the active nickel surface.[1, 11–14] The spacing between the carbon atoms of a double bond is close to 1·5 Å; between nickel atoms in their lattice 2·5 Å (Ref. 15) and copper about the same,[16] whilst the spacings of palladium, platinum and rhodium are all very close to 2·7 Å. The structure presented by these metal lattices is found to permit almost strain free association between them and carbon–carbon double bonds, the interval 2·7 Å being especially accommodating.[15–17] Further the exposed atoms at the edge and corners of a crystal, being the most exposed, could also be the most active. Thus inert supports are recognised as having a further useful effect in preventing crystallites from agglomerating or sintering during preparation. Even in an active catalyst only a minor proportion of the metal is engaged in catalytic activity. The marked catalytic activity of the metals mentioned above is very well known; it is obvious that some which are more active than nickel are too costly for normal industrial use.

Two rather similar mechanisms[18, 19] have been put forward to describe reactions at the nickel surface between hydrogen and carbon–carbon double bonds. It now appears from the detailed analysis of the hydrogenation of polyunsaturated fats that both are in simultaneous operation.[12, 13, 20, 21] Figure 4.5 illustrates schematically the situation where the 9C double bond of oleic acid having been adsorbed at the nickel surface (a), one hydrogen atom is next captured by carbon atom 9 (b).

According to the Horiuti–Polanyi mechanism there are now five possible reactions:

(i) The captured hydrogen atom is lost, the fatty acid chain di-

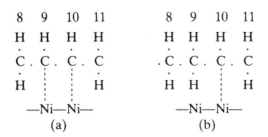

FIG. 4.5. Linkage of a double bond to catalyst atoms.

ssociates and we are back where we began with a 9C *cis* double bond.

(ii) The other 9C hydrogen atom is lost, the fatty acid chain dissociates but this time at 9C is a *trans* double bond.

(iii) A hydrogen is lost from 11C, the chain dissociates with a *cis* double bond at 10C.

(iv) The other hydrogen on 11C is lost, the chain dissociates with a *trans* double bond at 10C.

(v) 10C captures a hydrogen whilst 9C holds two, hence the bond is saturated and dissociates.

In the above (ii) represents geometric isomerisation, (iii) positional isomerisation and migration and (iv) migration and geometric change. Had 9C remained linked to the nickel whilst 10C captured a hydrogen the same reasoning allows a migration of the double bond in the other direction to 8C, thus a *cis* or a *trans* double bond at 8C may form in addition to those listed above. The other mechanism envisages that when the double bond between carbon atoms 9 and 10 is adsorbed that a hydrogen atom may be lost from 8C or 11C. Then, temporarily, an allytic type complex where the three carbon atoms (nos. 8, 9 and 10 or 9, 10 and 11) are held to the nickel surface is formed. From that point on gains and then losses of a hydrogen atom from one or other of the possible isomers which form produce migration and isomerisation comparable with the Horiuti–Polanyi mechanism, but the pattern of migration is somewhat different and leads to some difference in the proportion of different isomers at later stages.[13]

Of special interest is the possibility that in a case such as linoleic acid (C18 *cis* 9 *cis* 12) the adsorption of one double bond on the nickel may be followed by the loss of one hydrogen from the adjacent reactive methylene group at 11C. 11C then forms a link to the nickel followed immediately by the remaining double bond, so that all five carbon atoms (9C–13C) are temporarily bonded. If one hydrogen atom is then captured by a carbon atom at either extreme (9C or 13C), a conjugated system forms with double bonds as shown:

$$\overset{9}{-CH_2}-CH=CH-CH=\overset{13}{CH}- \quad \text{or} \quad \overset{9}{-CH}=CH-CH=CH-\overset{13}{CH_2}-$$

This is a particularly reactive system. The mechanism of the hydrogenation reaction with copper catalyst appears of necessity to include the conversion of the skipped or methylene interrupted arrangement so common in natural fats (—CH=CH—CH$_2$—CH=CH—) to a conjugated system (—CH$_2$—CH=CH—CH=CH—) which then hydrogenates very readily to a monoene. This explains why copper is several times more selective than nickel in reducing polyunsaturates to monoenes and its failure to catalyse the hydrogenation of monoenes. It is less active than nickel, more vulnerable to poisoning, re-use is not practical, rigorous removal of traces of copper from the product is essential and the operation must be separate from nickel if the best selectivity is to be obtained.[1, 22] The possible sequences of migration, isomerisation and saturation of double bonds in the hydrogenation of polyunsaturated systems such as fish oils continues to be actively investigated by Ackman and others.[23] The continuous observation of the proportion of *trans* isomers being formed, with appropriate feed back is put forward as one of the instrumental aids in achieving the texture required by the time a pre-set IV has been reached. The infra-red method of monitoring compares the proportion of *trans* isomers present with that of methyl elaidate (100%). The samples under test must not have more than 5% conjugated double bond unsaturation remaining. Further, when di- and tri-unsaturated acids, part or all *trans*, are present, separate peaks or shoulders on the main absorption peak at 10·36 μm for isolated *trans* double bonds can appear. In these circumstances prior knowledge of what texture corresponds to the expected spectroscopic findings for the oil being hardened is needed and has to be interpreted empirically. The change in texture due to a *trans* double bond being in one location or another is amply illustrated by a simple example of four isomers of oleic acid (Fig. 4.6).

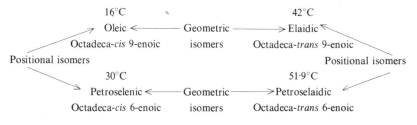

FIG. 4.6. The melting points of isomers of oleic acid.

When five or six double bonds in the fatty acid chain are present at the beginning of hydrogenation/migration/isomerisation the empiric nature of the control is evident.

4.5 REFERENCES

1. PATTERSON, H. B. W. (1983). *Hydrogenation of Fats and Oils*. Elsevier Applied Science, London.
2. Flow Sheet SFS – 164, EMI Corp., Des Plaines, USA.
3. Semi-continuous hydrogenation system, Technical Handout, Sullivan Systems Inc., Tiburon, USA.
4. Hydrogenation plant with post treatment, Technical Handout, Alfa-Laval AB, Tumba, Sweden.
5. Wurster & Sanger, Technical Handout, 222 W. Adams St., Chicago, USA.
6. Simon–Rosedowns Ltd, Technical Handout, Cannon St., Hull, UK.
7. S. A. De Smet Engineering N. V., Technical Handout, Antwerp, Belgium.
8. Lurgi Express Information, T1186/2.76. Frankfurt a. M., FRG.
9. (i) LEUTERITZ, G. (1969). Die kontinuierliche Hydrierung von Ölen und Fetten. *Fette Seif. Ans.*, **71**, 441. Also Technical Handout, BUSS AG., Basle, Switzerland.
   (ii) DUVEEN, R. F. and LEUTERITZ, G. (1982). Der BUSS-Schleifenreaktor in der Öl und Fetthärtungsindustrie, *ibid.*, **84**, 511.
10. Lightning Mixers Ltd Technical Handout, Stockport, UK.
11. BECKMANN, H. J. (1983). Technical Handout, Harshaw Chemie BV, De Meern, Netherlands and *JAOCS*, **60**, 282.
12. COENEN, J. W. E. (1978). The rate of change in the perspective of time. Sixth Hilditch Memorial Lecture. *Chem. Ind.*, 18 Sep., 709.
13. COENEN, J. W. E. (1970). *Margarine Today*. International Federation of Margarine Associations, E. J. Brill, Leiden, Netherlands.
14. BOLDINGH, J. (1969). Research. In *Margarine—An Economic, Social and Scientific History, 1869–1969*. Ed. J. H. van Stuyvenberg, Liverpool University Press, 194–200.
15. WATERMAN, H. I. (1951). *Hydrogenation of Fatty Oils*, Elsevier Applied Science, London.
16. WELLS, A. F. (1962, 1975). *Structural Inorganic Chemistry*. Clarendon Press, Oxford, 881, 879.
17. TWIGG, G. H. and RIDEAL, E. K. (1940). The chemisorption of olefines on nickel. *Trans. Faraday Soc.*, **36** 533.
18. HORIUTI, J. and POLANYI, M. (1934). *Trans. Faraday Soc.*, **30**, 396.
19. ROONEY, J. J., GAULT, F. G. and KEMBALL, C. (1960). *Proc. Chem. Soc.*, 407.
20. HEERTJE, I., KOCH, G. K. and WOSTEN, W. J. (1974). *J. Catal.*, **32**, 337, 1974.
21. VAN DER PLANK, P. and VAN OOSTEN, H. J. (1975). *ibid.*, **38**, 223.
22. DUTTON, H. J. (1982). Hydrogenated fats: processing, analysis and biological implications. Lewkowitsch Memorial Lecture. *Chem. Ind.*, 2 Jan., 9.
23. SEBEDIO, J. L., LANGMAN, M. F., EATON, C. A. and ACKMAN, R. G. (1981). Alteration of long chain fatty acids of herring oil during hydrogenation on nickel catalyst. *JAOCS*, **58**, 41.

5

The Analysis of Lipids with Special Reference to Milk Fat

WILLIAM W. CHRISTIE

*The Hannah Research Institute,
Ayr, Scotland, UK*

5.1 INTRODUCTION

Lipids in milk provide a major source of energy and essential structural components for the cell membranes in the tissues of the newborn in all mammalian species. They also fulfil these functions when supplied as constituents of other foodstuffs, and can confer distinctive properties on dairy foods that affect processing. For such reasons and because of the commercial importance of milk, milk lipids have probably been studied more intensively than those from any other source. The composition, structures and chemistry of milk lipids have been reviewed on a number of occasions.[1-5] Only 30 years ago, lipids were considered to be oily intractable substances that could be separated into simpler components only with difficulty. This picture was changed by the development of chromatographic procedures, initially gas–liquid chromatography (GLC) and thin-layer chromatography (TLC), and more recently high-performance liquid chromatography (HPLC). These, together with advances in spectroscopic methods, especially mass spectrometry, have brought about the explosive growth in knowledge of lipids in general and milk lipids in particular. Lipid analytical methodology has been reviewed.[6-9]

There are essentially four basic principles that govern the types of separation of lipids that can be achieved by chromatography, i.e. partition, adsorption, ion-exchange and complexation. GLC is the best known example of partition chromatography, and has been used to good effect for the analysis of the fatty acid constituents of milk lipids, but has also proved useful for other non-polar aliphatic residues and for molecular species of intact lipids. TLC is the main form of adsorption chromatography in common use, for example, for the separation of individual simple and complex lipid classes. Although ion-exchange chromatography is not used very often with lipids, there are some applications to the analysis of complex lipids, where distinctive separations have been achieved. Complexation chromatography should not perhaps be considered apart from the other techniques as it is always used in conjunction with one or other. As an example, silver nitrate-impregnated adsorbents are used to effect separations of lipids according to the number of double bonds in the molecules. HPLC can also be used in various modes, but mainly adsorption and reversed-phase partition, to achieve separations, and in essence uses the same principles as other methods. However, it utilises recent technological advances in instrumentation and in the preparation of microparticulate adsorbents and liquid phases to improve the resolution that can be obtained, and to introduce a degree of automation into the detection systems, the recording of separations and the handling of analytical data.

Milk lipid analysis has been a proving ground for many of these techniques. In this review, the application of chromatographic procedures will be considered in four areas of the analysis of milk lipids, i.e. the separation and quantification of lipid classes, the components of lipid classes (e.g. fatty acids), the positional distributions of fatty acids in glycerolipids and molecular species of glycerolipids. Information obtained in this way may be of great value for the nutritional evaluation of milk products, and to determine those aspects of the structures of milk lipids that relate to particular physical properties and consumer applications.

A large number of papers dealing with the analysis of milk lipids are published every year and it is not possible to review all of these comprehensively here. Rather, that work which in my opinion is definitive will be described together with other analyses which may not be unique but appear to be particularly good examples to illustrate specific points.

5.2 LIPID CLASS SEPARATIONS

The triacylglycerols are by far the major lipid class in milk, comprising 97–98% of the total, and they are accompanied by small amounts only of di- and monoacylglycerols, free cholesterol and cholesterol esters (in the

TABLE 5.1
COMPOSITION OF INDIVIDUAL LIPIDS IN BOVINE MILK[a]

Lipid class	Amount (wt %)	Lipid class	Amount (wt %)
Triacylglycerols	97·500	Phosphatidylcholine	0·207
Diacylglycerols	0·360	Phosphatidylethanolamine	0·191
Monoacylglycerols	0·027	Phosphatidylserine	0·019
Cholesterol esters	trace	Phosphatidylinositol	0·028
Cholesterol	0·310	Sphingomyelin	0·151
Free fatty acids	0·027	Lysophospholipids	trace

[a]Adapted from Ref. 5.

approximate ratio 10:1), unesterified (free) fatty acids and phospholipids as shown in Table 5.1. Comparatively large amounts of partial glycerides and unesterified fatty acids have been reported on some occasions, but this usually means that faulty handling of the milk has led to some lipolysis. In most circumstances, TLC would be the method of choice for the analysis of these compounds;[8] they are separated rather easily, but the small amounts of most relative to the triacylglycerols tends to lead to problems in quantification. Charring of the lipids with a corrosive spray followed by photodensitometry of the charred spots is the method used for quantification in most applications, but may not be sufficiently sensitive for the minor constituents of milk fats. A more accurate procedure consists in transesterifying each of the separated lipids in the presence of a suitable internal standard, generally a fatty acid not found naturally in the sample, and then quantifying the resulting methyl ester derivatives by GLC.[10]

More distinctive methods are available for specific simple lipid components. For example, it may be important to monitor lipolysis in milk by measuring the free fatty acid content and composition, and a GLC method has been devised for the purpose.[11] Other sensitive enzymic[12,13] and radiochemical[14] methods have been described but have yet to be applied to dairy products. A new HPLC procedure appears to be particularly useful;[15] the free acids are converted to the p-bromophenacyl esters, via a crown ether-catalysed reaction, without separation from other butter components, and are separated and quantified on a reversed-phase (C_{18}-bonded) column with UV detection.

Lipids containing free hydroxyl groups, present in milk, have been determined by a sensitive spectrophotometric assay following conversion to the pyruvic ester-2,6-dinitrophenylhydrazone derivatives;[16] they include diacylglycerols, hydroxyacylglycerols and sterols. In addition, cholesterol can be determined by a variety of chromatographic and enzymic methods,[8] while sterols other than cholesterol in milk have been identified and quantified by GLC.[17-20] Sensitive radio-immunological methods must be used for steroid hormones,[21] and for prostaglandins,[22] although ideally they should be used in concert with chromatographic procedures.

The complex lipids in milk have been fractionated by methods analogous to those used for the same lipid classes in animal tissues in general. Phospholipids, for example, can be fractionated by TLC procedures,[8] and related methods have been used for glycolipids including gangliosides,[23-24] although HPLC methods that appear simpler and have greater sensitivity are available for the latter.[8]

Great strides are being made in the development of HPLC methods for the analysis of most lipid classes, and some procedures for the analysis of phospholipids appear particularly promising. A considerable effort has been expended in the search for a lipid-sensitive detector for monitoring the eluate from HPLC columns. Differential refractometers, which sense minute differences in the refractive index of the eluate brought about by compounds eluting from the columns, have been used but are not suitable for gradient-elution applications. UV detectors, operating at 200–206 nm, have also been applied to the problem but can only be used with a limited range of solvents and these must be of the highest purity. A system in which the column-eluate was coated onto a moving-wire and passed through a flame-ionisation detector gave excellent results in some hands,[25-28] but was not a commercial success initially, although an instrument that uses this principle is again available. A new detector that appears to offer a number of advantages is the 'mass-detector', which utilises light-scattering following evaporation of the solvent in a stream of compressed air, as the detection principle;[29] it is available commercially from Applied Chromatography Systems Ltd (Luton, Beds).

An application of this detector in the analysis of milk phospholipids is illustrated in Fig. 5.1 (Author, unpublished work). As these components represent a rather small fraction of the total lipids in milk, they were first concentrated by a solvent partition procedure.[30] The column contained Spherisorb™ 3μ silica and was eluted first with an isooctane–isopropanol gradient to elute the residual simple lipids and then with a gradient of water into isopropanol–hexane to elute the individual phospholipids. The procedure is rapid and quantitative, if calibrated carefully. At present, it appears that the main drawback is that the detector

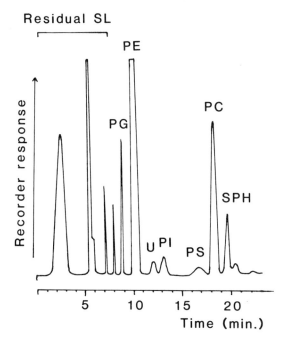

FIG. 5.1. HPLC separation of a phospholipid fraction prepared from milk lipids (author, unpublished work). The column (10 cm × 0·5 cm) was packed with Spherisorb™ 3μ silica and the 'mass detector' (Applied Chromatography Systems, Luton, Beds) was used with a Spectra-Physics ternary solvent delivery system (Spectra-Physics Ltd, St. Albans, Herts). A gradient of isopropanol into isooctane was used in the first 5 min to elute the residual simple lipids, and a gradient of water into isopropanol–hexane was used over the next 15 min to elute the phospholipids; the flow rate was 2 ml/min. Abbreviations: SL, simple lipids; PG, phosphatidylglycerol; PE, phosphatidylethanolamine; U, unknown; PI, phosphatidylinositol; PS, phosphatidylserine; PC, phosphatidylcholine; SPH, sphingomyelin.

response is not linearly proportional to the amount of material eluting but tends to fall off somewhat at lower concentrations. It is obviously a method that will be much more widely used in future.

5.3 FATTY ACIDS AND RELATED ALIPHATIC COMPOUNDS

A greater range of fatty acids has been isolated or identified as components of milk fats than from any other natural source. For example, a

TABLE 5.2
THE NUMBERS OF FATTY ACIDS OF VARIOUS TYPES FOUND IN BOVINE MILK[a]

Type	Number
Normal saturated	27
Mono-branched	71
Multi-branched	18
Cis-monoenes	57
Trans-monoenes	58
Dienes	42
Polyenes	18
Keto	59
Hydroxy	85
Cyclic	2
Total	437

[a] Adapted from Patton and Jensen.[2]

compilation of the fatty acids that had been detected in bovine milk by 1974 listed 437 distinct constituents;[2] they include all the odd- and even-numbered normal saturated fatty acids from C_2 to C_{28}, monomethyl-branched fatty acids from C_{11} to C_{28} (including positional isomers), multi-methyl-branched fatty acids, monoenoic fatty acids from C_{10} to C_{26} (115 configurational and positional isomers in total), a number of di- and poly-enoic fatty acids, and keto-, hydroxy- and cyclohexyl fatty acids. They are summarised in Table 5.2. Many of these are unique to ruminants and are products or by-products of biohydrogenation in the rumen, followed often by further metabolism in the tissues of the animal (e.g. by oxidation, chain-elongation or desaturation). With non-ruminants, fewer distinct fatty acids would be expected in the milk, but 183 different components have been identified so far in human milk.[31] Most of these fatty acids have only been detected by using combinations of the more advanced chromatographic, chemical and spectroscopic techniques available. Fortunately, a relative few of these need to be determined when making nutritional evaluations of milk and dairy foods.

GLC is the single most valuable method in use for the analysis of the fatty acid constituents of milk. For this purpose, it is first necessary to transesterify the lipids to more volatile derivatives such as the methyl esters. Because of the volatility and partial solubility in water of the short-chain esters derived from milk fat, the best methods are those with no aqueous-extraction or solvent-removal steps and where the reagents are not heated at any stage. An alkaline transesterification procedure using sodium methoxide in methanol, described by Christopherson and Glass,[32] best meets these criteria and has been widely adopted. A related method can be used for lipids in milk other than the triacylglycerols.[33] Although many research groups appear to get satisfactory results by transmethylation, it has been argued that more reproducible GLC analyses are obtained if the butyl ester derivatives of the fatty acids are prepared.[34]

With conventional packed-columns and polyester liquid phases in the gas chromatograph under optimum conditions, up to 53 distinct fatty acids were resolved, identified and quantified in milk fat in a single analysis, for example;[35] these included all the odd- and even-numbered normal fatty acids from 4:0 to 24:0, monoenoic fatty acids from 10:1 to 24:1, and a number of branched-chain and polyunsaturated fatty acids. With the greater degree of resolution attainable by capillary-column GLC, 80 distinct components of milk fat were described in one analysis (although silver-nitrate chromatography was also used here);[36] in addition to those fatty acids mentioned above, many positional isomers of the branched-chain and unsaturated fatty acids were resolved.

Recently, an HPLC method for the isolation of milk fatty acid derivatives on a semi-preparative scale has been described; it utilises a

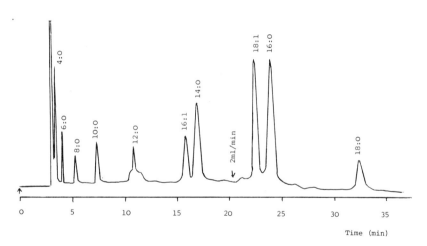

FIG. 5.2. Reversed-phase separation of milk butylesters on a column of Lichrosorb™ 10RP18 (25 cm × 0·5 cm) eluted with acetonitrile-water (98:2, v/v) at 1 ml/min initially then at 2 ml/min, with refractive index detection. Adapted from Christie et al.[37]

reversed-phase column (C_{18} bonded), isocratic elution with acetonitrile–water mixtures and a refractive-index detector (Fig. 5.2). It should be possible to develop a method, based on this principle, with aromatic esters and UV detection for quantitative analysis of milk fatty acids. There should then be no requirement to vaporise short-chain esters, as in gas chromatography, where losses are possible.[34, 38]

While the structure and nature of many of the minor fatty acid components might be considered to have only academic interest, revealing aspects of fatty acid metabolism in the mammary gland and other tissues, those of a few are of real significance. It is important to recognise, for example, that part of the component designated '18:2' or linoleic acid on analysis by GLC may consist of isomers other than cis-9, cis-12 and so may lack biological potency as an essential fatty acid. Indeed, it has been reported that only half the octadecadienoic acid in cows' milk may be the essential isomer.[39] Similarly, a proportion of the mono-unsaturated fatty acids comprises *trans*-isomers, the nutritional value of which has been the subject of some debate recently.[40] Methods for the analysis of such compounds are, therefore, of some importance.

TLC with silica gel layers impregnated with silver nitrate has been of particular value in the analysis of the minor fatty acid components. The separation is based on the property that silver salts form polar complexes reversibly with double bonds that retard the migration of the compounds on thin-layer adsorbents;[8, 41] saturated compounds do not form complexes so the developing solvent carries them ahead of compounds with one double bond, and these are in turn ahead of compounds with two double bonds and so forth. In addition, *trans*-double bonds do not form complexes as readily as *cis*-double bonds, so fatty acids can also be separated according to the configuration of their double bonds. Although model mixtures of *cis*- and *trans*-isomers have been separated by GLC on capillary and packed columns containing high-polarity liquid phases, the complexity of many fat samples, including milk fat is such that the technique can rarely be of practical value.[8]

Silver ions linked to sulphonic acid-based ion-exchangers and used in HPLC columns are also of considerable value for this purpose.[42–45] Rather polar solvents are required to elute the fatty acid methyl esters, but the silver ions do not leach from the columns, which can be used repeatedly.

Cis- and *trans*-monoene fractions isolated by silver nitrate TLC contain fatty acids differing greatly in chain length; a component of a given chain length can also contain many positional isomers. For further subdivision of monoene fractions according to chain length, preparative-scale GLC, which can be used with about 0·5 mg of material (more with special columns), has generally been the method of choice, although reversed-phase HPLC might now be favoured.[8] With the latter technique, there is no need to volatilise the sample and then condense it following separation for collection purposes, and there need be no losses in the detector.[8,46] It is then necessary to oxidise each of the chain length fractions, obtaining cleavage between the carbon atoms of the double bonds, and analyse the fragments by means of GLC to determine the positions of those double bonds. Ozonolysis or von Rudolff (permanganate–periodate) oxidation are here the methods of choice, as they cause a minimum of side reactions.[8,46,47] By using silver-nitrate TLC, preparative GLC and oxidative fission in sequences in this way, the component designated an '18:1' fatty acid in milk fat was found to consist of 4 cis- and 11 trans-isomers for example (Table 5.3).[48] Analogous methods have been used for the di- and poly- unsaturated fatty acids in milk,[49,50] and as mentioned earlier, this may be particularly important for the essential fatty acids.[39]

Useful separations of unsaturated isomers have been obtained by silver-nitrate TLC followed by capillary-column GLC, but fewer isomers can be quantified than when oxidative fission is used.[36,51]

GLC and mass spectrometry in combination can be a very powerful tool and have been used for the tentative identification of many of the

TABLE 5.3
OCTADECENOIC ACID ISOMERS IN MILK FAT (% OF THE TOTAL)[a]

Position of double bond	Cis-isomers	Trans-isomers
6		1·0
7		0·8
8	1·7	3·2
9	95·8	10·2
10	trace	10·5
11	2·5	35·7
12		4·1
13		10·5
14		9·0
15		6·8
16		7·5

[a] Adapted from Hay and Morrison.[48]

unsaturated fatty acids in milk fat.[36] While it reveals the numbers of double bonds in unsaturated fatty acids, it does not necessarily give information on their positions unless they are first converted to appropriate derivatives,[8] and no such application to milk fat appears to have been described. The technique has, however, been used to great advantage for the identification and analysis of the minor branched-chain fatty acids, present in bovine milk, that are not amenable to conventional chemical analysis.[52,53]

Aliphatic compounds contribute greatly to the flavour and palatability of milk and dairy products, and their compositions have been reviewed.[1,54,55] Very many different compounds are involved and a high proportion are derived, chemically or enzymatically, from milk lipids, e.g. by hydrolysis or oxidation. Those studied most intensively are lactones and methyl ketones, but short-chain aldehydes and fatty acids are also important. GLC coupled with mass spectrometry has proved particularly useful in their analysis.

The sphingolipids of cows' milk contain a complex range of aliphatic long-chain bases, derived biosynthetically from fatty acids, including normal, *iso*- and *anteiso*-saturated di- and trihydroxy isomers.[56–58] To analyse these compounds, they were first oxidised to remove the base moiety so that the non-polar aliphatic residues could be identified and quantified by GLC–mass spectrometry.

5.4 THE POSITIONAL DISTRIBUTIONS OF FATTY ACIDS IN MILK TRIACYLGLYCEROLS

Triacylglycerols are synthesised in the mammary gland by enzymic mechanisms that exert some specificity in the esterification of different fatty acids at each position of the L(or *sn*-)-glycerol moiety. Distinctive structures result that affect the physical properties and digestibility of milk fat. The composition of position *sn* 2 of the triacylglycerols is perhaps of greatest importance for the consumer, as during digestion by simple-stomached animals and by new-born ruminants, 2-monoacyl-*sn*-glycerols are formed by the action of the enzyme pancreatic lipase. It appears that such compounds containing relatively high proportions of palmitic acid, as do those from milk fat, have higher digestibility especially in the new-born.[59,60] On the other hand, the compositions of each of the positions is important to the biochemist and may be relevant to the physical properties in consumer applications of milk fat.

The composition of position *sn* 2 of milk triacylglycerols can be determined by means of hydrolysis *in vitro* with pancreatic lipase, an enzyme which is almost entirely specific for the primary esters of triacylglycerols. 2-Monoacyl-*sn*-glycerols are formed that can be separated from the other products of the reaction by TLC before the fatty acid constituents are analysed by GLC.[8,61] For some time, it was thought that there was preferential hydrolysis of the short-chain fatty acids, but it is now known that this is not so although there is indeed some more rapid hydrolysis of those molecular species of triacylglycerols containing short-chain fatty acids.[62] In practice, the procedure appears to give representative monoacylglycerols and, therefore, satisfactory results.

No lipolytic enzyme has yet been discovered that distinguishes between positions 1 and 3 of a triacyl-*sn*-glycerol, but a number of ingenious, if complicated, stereospecific analysis procedures have been devised and these have been reviewed.[8,63] Most of the available methods require the preparation of diacylglycerols, which are then chemically phosphorylated for reaction with a stereospecific enzyme, or which are phosphorylated by stereospecific enzymes. The author[8] favours an approach in which α,β-diacylglycerols (an equimolar mixture of the 1,2- and 2,3-*sn*-isomers) are prepared by partial hydrolysis of the triacylglycerols with a Grignard reagent, and are converted synthetically to phosphatidylcholines; these are in turn reacted with the phospholipase A of snake venom, which only hydrolyses the 'natural' 1,2-diacyl-*sn*-glycerophosphorylcholine. The products are lysophosphatidylcholine, which contains the fatty acids originally present in position *sn* 1, free fatty acids released from position *sn* 2 and the unchanged 2,3-diacyl-*sn*-glycerophosphorylcholine. Each of these products is isolated by means of TLC and is transesterified for analysis of the fatty acid constituents by GLC. This procedure borrows elements from the work of several research groups.[64-66] The final separation step in the procedure can now be accomplished more tidily by means of HPLC on a silica column, prior to GC analysis.[67]

In the analysis of milk lipids by this procedure, problems are encountered in the preparation of pure α,β-diacylglycerols, because the wide range of chain lengths of the fatty acid constituents causes band spreading during isolation by means of TLC, and some contamination with 1,3-diacyl-*sn*-glycerols results. The first moderately successful approach[68] utilised α,β-diacylglycerols prepared by means of pancreatic lipase hydrolysis, but they may not have been entirely representative of those in the triacylglycerols because of selective hydrolysis of the lower molecular weight species. Others preferred to separate ruminant milk fats by TLC or molecular distillation into a fraction containing predominantly the long-chain fatty acids and one containing the short-chain fatty acids, and to subject these independently to stereospecific analysis, the results being combined later.[69-72] It is also possible, if somewhat less accurate, to use a similar approach, but combine the end products from hydrolysis of the long-chain and short-chain fractions for fatty acid analysis.[73]

The results of such analyses show that there is a clear preference for palmitic acid to be esterified to position sn 2 of the triacylglycerols, not only in the milk of ruminant animals but also of virtually all other mammalian species.[5,74,75] In ruminant milks, a further distinctive feature is that the short-chain fatty acids, especially butyric and hexanoic acids, are concentrated entirely in position sn 3.[68-73]

The positional distributions of the fatty acids in the main glycerophospholipid constituents of milk, i.e. phosphatidylcholine and phosphatidylethanolamine, have been determined by means of phospholipase A hydrolysis; these lipids are also asymmetric, but not as markedly so as those in many other tissues of the animal.[76,77]

5.5 MOLECULAR SPECIES OF MILK TRIACYLGLYCEROLS

The complete description of a natural triacylglycerol requires that it be separated into single molecular species in which each position of the glycerol moiety is occupied by only one fatty acid. Patton and Jensen[2] have pointed out that the complexity of milk fatty acids is such that 64×10^6 individual triacylglycerol species could be present, 'the identification of which would keep generations of biochemical taxonomists busy and happy'. Even if only the 20 or so most abundant fatty acids are considered, there could be 8000 molecular species including all positional (and enantiomeric) isomers. It is only realistic, therefore, to attempt to isolate relatively simple molecular fractions rather than single species.

Crude separations into high and low molecular weight fractions are possible by molecular distillation, low-temperature crystallisation and adsorption chromatography. While such procedures may have some practical applications, they tend to have comparatively little analytical value and are not considered further here.

From a practical standpoint, the most valuable analytical procedures are those that enable a rapid 'fingerprint' of triacylglycerol molecular species to be obtained so that the physical properties, such as melting or softening points, for example, can be related to definable compositional or structural features. High-temperature GLC can be used for this purpose, and HPLC methods are under development that hold promise for the future.

In high-temperature GLC analyses of triacylglycerols, separation is normally achieved into fractions with the same combined molecular weights of their fatty acid constituents; the combined chain lengths of the fatty acyl groups is termed the 'carbon number' of the triacylglycerol.[8] No separation according to degree of unsaturation is achieved with packed columns. Commonly, short glass columns with low levels of non-polar silicone liquid phases are used, as the temperatures (up to 350°C) required to elute the higher molecular weight constituents approach the limits of thermal stability both of the stationary phases and of the compounds themselves. Milk fat fractions differing by two carbon atoms from carbon number 30 to 54 have been successfully separated by this means.[71,78] Unfortunately, the presence of odd-chain and branched-chain fatty acids in milk leads to some peak broadening and loss of resolution. Although there is little margin for error in the preparation of the columns, and some preliminary work and practice may be necessary to achieve the required separations and to calibrate for quantitative analysis, this technique appears to be the best available for routine analysis of large numbers of samples.

High-temperature GLC with capillary columns (with glass or fused-silica walls) affords much greater resolution and some preliminary applications to milk triacylglycerols have been described, Components containing odd-chain and branched-chain fatty acids are clearly resolved, and some separation according to degree of unsaturation also appears to be achieved, even with non-polar silicone phases.[79-83] So many components are resolved by this technique indeed, that individual peaks on the recorder trace are not readily identifiable unless access is available to a mass spectrometer.[82,84] Some technical problems remain in obtaining accurate quantification,[79-81,85,86] but the technique will undoubtedly be developed further.

Excellent separations of triacylglycerols have been achieved by means of HPLC in the reversed-phase mode, and some applications have been reviewed.[8,87-90] By far the most widely used stationary phase consists of octadecyl groups covalently-bound onto microparticles of silica, 3–10 μm in diameter, and it is available under a number of trade names. A number of detection systems have been used (see Section 5.2 above). Excellent results in lipid applications have been obtained by using a mass spectrometer interfaced to an HPLC system as detector,[91-94] but such equipment is probably too costly for routine use in many laboratories. It does, however, offer the very considerable advantage that fractions are not only quantified but that the fatty acid compositions are simultaneously determined. With a complex mixture, such as milk fat, this could be of great assistance to the analyst. Infra-red spectroscopy detectors (at 5·75 µm for the carboxyl function) have been used to some extent although only a few applications to lipids have been described to date.[95-97] A few applications of flame ionisation detectors to triacylglycerol analyses have also been reported.[87-89, 98]

The 'mass detector' is being used increasingly for the analysis of molecular species of triacylglycerols, including those of milk fat.[99-102] One such application is illustrated in Fig. 5.3 (author, unpublished work), where cows' milk triacylglycerols have been separated by reversed-phase chromatography with a gradient of acetone in acetonitrile. A large number of components are clearly resolved by this procedure

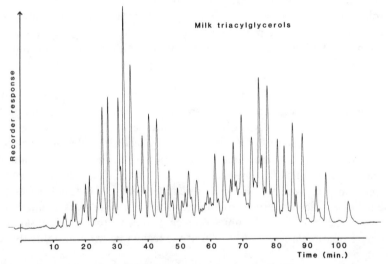

FIG. 5.3. Reversed-phase separation of cows' milk triacylglycerols by reversed-phase HPLC on a column of Spherisorb™ 5µ ODS2 (25 cm × 0·5 cm) with mass detection (author, unpublished work). The eluting solvent was acetone–acetonitrile (1:1, v/v) for the first 15 min and then was changed by a linear gradient to acetone (100%) over a further 105 min; the flow rate was 0·5 ml/min. The detector and HPLC equipment was as Fig. 5.1.

Hydrogenated milk triacylglycerols

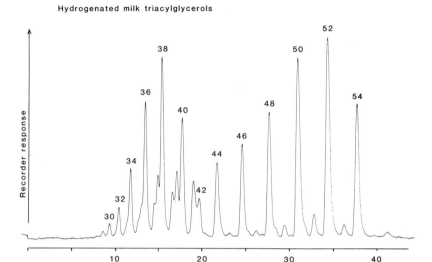

FIG. 5.4. Reversed-phase separation of cow's milk triacylglycerols following hydrogenation (author, unpublished work). Column and HPLC equipment as Fig. 5.3. The eluting solvent was dichloroethane–acetonitrile (1:1, v/v) for 5 min and was changed via a linear gradient to dichloroethane–acetonitrile (4:1, v/v) over a further 55 min. The numbers above peaks refer to the total numbers of carbon atoms in the fatty acid constituents.

and many more can be seen as 'shoulders' on peaks; further resolution would undoubtedly be possible by slowing the gradient or increasing the length of the column, for example. Molecular species of triacylglycerols are separated by reversed-phase HPLC according to their partition number, where partition number = carbon number − 2 × (number of double bonds), i.e. a double bond tends to reduce the partition number by 2 units; as separation efficiencies have improved with technical developments, this factor can sometimes be greater than 2. It is not easy, therefore, to interpret the chromatogram shown in Fig. 5.3 in molecular terms. Even when the chromatograph is simplified by hydrogenating the milk fat prior to analysis (as in Fig. 5.4), the pattern obtained is still rather complex because of the presence of odd-chain, branched-chain and short-chain fatty acids in milk fat. A dichloroethane–acetonitrile gradient was used for elution in this instance because of the lack of solubility of the saturated high molecular weight triacylglycerols in acetone–acetonitrile mixtures. Again by optimising the separation para-

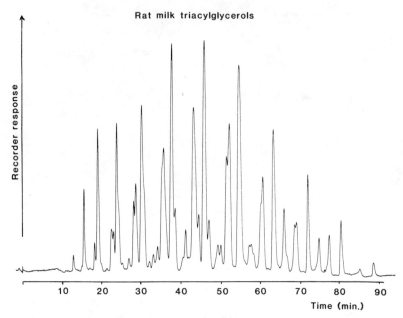

FIG. 5.5. Reversed-phase separation of rat milk triacylglycerols by means of HPLC (author, unpublished work). Column, HPLC equipment and elution scheme as in Fig. 5.3.

meters, a great improvement in resolution was possible, and as many as 4 distinct components could be clearly resolved between those triacylglycerol species differing in carbon number by only 2 units (author, unpublished work). These separations then appear to be an elegant demonstration of the resolving power of reversed-phase HPLC, but are unlikely to be of value for routine analytical purposes with cows' milk. The technique may prove useful nonetheless in biochemical experiments with species other than ruminants. For example, with rat milk, where odd-chain and branched-chain components are not present, some useful separations are possible (Fig. 5.5).

The chromatographs illustrated in Figs. 5.3, 5.4 and 5.5 indicate the value of the mass detector for qualitative purposes. It is to be hoped that data will soon be available on the quantitative value of the procedure. For example, it is already evident that the response of the detector is affected by instrumental parameters, such as the flow rate and temperature of the gas in the evaporator unit, and the nature and flow rate of the eluting solvent. The nature of the lipids separated, for example their structures, molecular weight and degree of unsaturation, may also be important.

In addition to the analytical separations, there is a need for a larger scale method so that sufficient material can be obtained in each molecular fraction for analysis by other methods, and for determination of physical properties, such as by differential scanning calorimetry, for example. The most practicable approach uses HPLC technology adapted to a preparative scale (also in a reversed-phase mode) with (C_{15}–C_{18})-alkoxypropyl-Sephadex LH20 as the stationary phase.[103]

For some purposes, it may be necessary to obtain more information on the molecular structure of a milk fat than can be obtained by analysis by one of the techniques described above. Combinations of methods that make use of different separatory principles, must then be used in sequence together perhaps with stereospecific analysis procedures. The most common combination consists in essence of silver-nitrate chromatography(TLC) and high-temperature GLC, and a detailed protocol for the analysis of milk fat in this way has been described.[71] In silver-nitrate chromatography in this instance, separation is achieved according to the total number of double bonds in the three acyl moieties, i.e. molecular species with three saturated fatty acids migrate ahead of those with two saturated and one monoenoic fatty acid, and so forth. HPLC separations of triacylglycerols on silica impregnated with silver nitrate have also been described but do not appear to have been applied to milk fat.[87-89, 104] Unfortunately, silver-impregnated ion-exchange resins could not be used with triacylglycerols as some residual sulphonic acid residues caused interesterification to occur as a side reaction.[45]

Another time-consuming but fruitful approach has been reversed-phase column chromatography and pancreatic lipase hydrolysis of the fractions, by which means 168 molecular species of the shorter chain triacylglycerols of milk fat were isolated and quantified.[105] Related methods have been used by many research groups for the analysis of milk triacylglycerols.[106-115]

5.6 CONCLUSIONS

Analysts have stretched many modern chromatographic methods to their limits in the analysis of milk lipids. In consequence, we know more of the detailed structures of milk lipid constituents than of the lipids from any other natural source. The techniques of capillary GLC and HPLC are being developed particularly rapidly, and hold promise in terms of speed of analysis, resolution and convenience for the future.

5.7 ACKNOWLEDGEMENT

A large part of the text of this paper was published earlier in 'Challenges to Contemporary Dairy Analytical Techniques, Royal Society of Chemistry, Special Publication No. 49' (1984) pp. 139–154. The author, editors, and publisher are pleased to acknowledge that the Royal Society of Chemistry kindly gave permission for the reproduction.

5.8 REFERENCES

1. MORRISON, W. R. (1970). *Topics in Lipid Chem.*, **1**, 51–106.
2. PATTON, S. and JENSEN, R. G. (1975). *Prog. Chem. Fats*, **14**, 163–277.
3. KURTZ, F. E. (1974). In *Fundamentals of Dairy Chemistry*. 2nd edn. Ed. B. H. Webb, A. H. Johnson, and J. A. Alford. AVI Publishing Co., Westport, 125–219.
4. CHRISTIE, W. W. (1981). *Lipid Metabolism in Ruminant Animals*. Ed. W. W. Christie. Pergamon Press, Oxford, 95–191.
5. CHRISTIE, W. W. (1983). In *Developments in Dairy Chemistry. 2. Lipids*. Ed. P. F. Fox. Elsevier Applied Science, London, 1–35.
6. CHRISTIE, W. W. (1980). In *Fats and Oils: Chemistry and Technology*. Ed. R. J. Hamilton, and A. Bhati. Elsevier Applied Science, London, 1–23.
7. KUKSIS, A. (1978). *Handbook of Lipid Research. Vol. 1*. Plenum Press, New York.
8. CHRISTIE, W. W. (1982). *Lipid Analysis*. Pergamon Press, Oxford.
9. CHRISTIE, W. W. and NOBLE, R. C. (1984). In *Food Constituents and Food Residues: Their Chromatographic Determination*. Ed. J. F. Lawrence. Marcel Dekker Inc., New York, 1–50.
10. CHRISTIE, W. W., NOBLE, R. C. and MOORE, J. H. (1970). *Analyst (London)*, **95**, 940–4.
11. DEETH, H. C., FITZ-GERALD, C. H. and SNOW, A. J. (1983). *N.Z.J. Dairy Sci. Technol.*, **18**, 13–20.
12. MIZUNO, K., TOYOSATO, M., YABUMOTO, S., TANIMIZU, I. and HIRAKAWA, H. (1980). *Analyt. Biochem.*, **108**, 6–10.
13. SHIMIZU, S., INOUE, K., TANI, Y. and YAMADA, H. (1979). *Analyt. Biochem.*, **98**, 341–5.
14. DE BRABANDER, H. F. and VERBEKE, R. (1981). *Analyt. Biochem.*, **110**, 240–1.
15. REED, A. W., DEETH, H. C. and CLEGG, D. E. (1984). *J. Assoc. Off. Anal. Chem.*, **67**, 718–21.
16. TIMMEN, H. and DIMICK, P. S. (1972). *J. Dairy Sci.*, **55**, 919–25.

17. BREWINGTON, D. R., CARESS, E. A. and SCHWARTZ, D. P. (1970). *J. Lipid Res.*, **11**, 355–61.
18. FLANAGAN, V. P., FERRETTI, A., SCHWARTZ, D. P. and RUTH, J. M. (1975). *J. Lipid Res.*, **16**, 97–101.
19. ADACHI, A. and KOBAYASHI, T. (1979). *J. Nutrit. Sci. Vitaminol.*, **25**, 67–78.
20. PARODI, P. W. (1973). *Austral. J. Dairy Technol.*, **28**, 135–7.
21. POPE, G. S. and SWINBURNE, J. K. (1980). *J. Dairy Res.*, **47**, 427–49.
22. MATERIA, A., JAFFE, B. M., MONEY, S. R., ROSSI, P., DE MARCO, M. and BASSO, N. (1984). *Arch. Surg.*, **119**, 290–2.
23. HUANG, R. T. C. (1973). *Biochim. Biophys. Acta*, **306**, 82–4.
24. KEENAN, T. W. (1974). *Biochim. Biophys. Acta*, **337**, 255–70.
25. STOLYHWO, A. and PRIVETT, O. S. (1973). *J. Chromatogr. Sci.*, **11**, 20–5.
26. STOLYHWO, A., PRIVETT, O. S. and ERDAHL, W. L. (1973). *J. Chromatogr. Sci.*, **11**, 263–7.
27. PHILLIPS, F. C. and PRIVETT, O. S. (1981). *J. Am. Oil Chemists Soc.*, **58**, 590–4.
28. PHILLIPS, F. C., ERDAHL, W. L. and PRIVETT, O. S. (1982). *Lipids*, **17**, 992–7.
29. CHARLESWORTH, J. M. (1978). *Analyt. Chem.*, **50**, 1414–20.
30. GALANOS, D. S. and KAPOULAS, V. M. (1962). *J. Lipid Res.*, **3**, 134–6.
31. JENSEN, R. G., CLARK, R. M. and FERRIS, A. M. (1980). *Lipids*, **15**, 345–55.
32. CHRISTOPHERSON, S. W. and GLASS, R. L. (1969). *J. Dairy Sci.*, **52**, 1289–90.
33. CHRISTIE, W. W. (1982). *J. Lipid Res.*, **23**, 1072–5.
34. IVERSON, J. L. and SHEPPARD, A. J. (1977). *J. Assoc. Off. Anal. Chem.*, **60**, 284–8.
35. MELCHER, F. and RENNER, E. (1976). *Milchwissenschaft*, **31**, 70–6.
36. STROCCHI, A. and HOLMAN, R. T. (1971). *Riv. Ital. Sostanze Grasse*, **48**, 617–22.
37. CHRISTIE, W. W., CONNOR, K. and NOBLE, R. C. (1984). *J. Chromatogr.*, **298**, 513–515.
38. KIRK, R. S. (1984). *Challenges to Contemporary Dairy Analytical Techniques*. Royal Society of Chemistry, London, 127–31.
39. KIURU, K., LEPPANEN, R. and ANTILA, M. (1974). *Fette Seifen Anstrichm.*, **76**, 401–8.
40. KINSELLA, J. E., BRUCKNER, G., MAI, J. and SHIMP, J. (1981). *Am. J. Clin. Nutr.*, **34**, 2307–18.
41. MORRIS, L. J. (1966). *J. Lipid Res.*, **7**, 717–32.
42. HOUX, N. W. H. and VOERMAN, S. (1976). *J. Chromatogr.*, **129**, 456–9.
43. SCHOFIELD, C. R. (1979). In *Geometrical and Positional Fatty Acid Isomers*. Ed. E. A. Emken and H. J. Dutton. American Oil Chemists Society, Champaign, Ill., 17–52.
44. ADLOF, R. O., RAKOFF, H. and EMKEN, E. A. (1980). *J. Am. Oil Chem. Soc.*, **57**, 273–5.
45. ADLOF, R. O. and EMKEN, E. A. (1980). *J. Am. Oil Chem. Soc.*, **57**, 276–8.
46. SEBEDIO, J. L., FARQUHARSON, T. E. and ACKMAN, R. G. (1982). *Lipids*, **17**, 469–75.
47. DUTTON, H. J. (1975). *Analysis of Lipids and Lipoproteins*. Ed E. G. Perkins. American Oil Chemists' Society, Champaign, Ill., 138–52.

48. HAY, J. D. and MORRISON, W. R. (1970). *Biochem. Biophys. Acta*, **202**, 237–43.
49. JENSEN, R. G., QUINN, J. G., CARPENTER, D. L. and SAMPUGNA, J. (1967). *J. Dairy Sci.*, **50**, 119–26.
50. VAN DER WEL, H. and DE JONG, K. (1969). *Fette Seifen Anstrichm.*, **67**, 279–81.
51. LUND, P. and JENSEN, F. (1983). *Milchwissenschaft*, **38**, 193–6.
52. MASSART-LEEN, A. M., DE POOTER, H., DECLOEDT, M. and SCHAMP, N. (1981). *Lipids*, **16**, 286–92.
53. LOUGH, A. K. (1977). *Lipids*, **12**, 115–19.
54. FORSS, D. A. (1971). *J. Am. Oil Chem. Soc.*, **48**, 702–10.
55. FORSS, D. A. (1972). *Prog. Chem. Fats*, **13**, 177–258.
56. MORRISON, W. R. and HAY, J. D. (1970). *Biochim. Biophys. Acta*, **202**, 460–7.
57. MORRISON, W. R. (1971). *FEBS Letters*, **19**, 63–4.
58. MORRISON, W. R. (1973). *Biochim. Biophys. Acta*, **316**, 98–107.
59. TOMARELLI, R. M., MEYER, B. J., WEABER, J. R. and BERNHART, F. W. (1968). *J. Nutr.*, **95**, 583–90.
60. FILER, L. J., MATTSON, F. H. and FOMON, S. J. (1969). *J. Nutr.*, **99**, 293–8.
61. BROCKERHOFF, J. and JENSEN, R. G. (1974). *Lipolytic Enzymes*. Academic Press, New York.
62. SAMPUGNA, J., QUINN, J. G., PITAS, R. E., CARPENTER, D. L. and JENSEN, J. G. (1967). *Lipids*, **2**, 397–402.
63. BRECKENRIDGE, W. C. (1978). *Handbook of Lipid Research, Vol. 1*. Ed. A. Kuksis. Plenum Press, New York, 197–232.
64. BROCKERHOFF, H. (1965). *J. Lipid Res.*, **6**, 10–15.
65. CHRISTIE, W. W. and MOORE, J. H. (1969). *Biochim. Biophys. Acta*, **176**, 445–52.
66. MYHER, J. J. and KUKSIS, A. (1979). *Can. J. Biochem.*, **57**, 117–24.
67. CHRISTIE, W. W. and HUNTER, M. L. (1984). *J. Chromatogr.*, **294**, 489–93.
68. PITAS, R. E., SAMPUGNA, J. and JENSEN, R. G. (1967). *J. Dairy Sci.*, **50**, 1332–6.
69. BRECKENRIDGE, W. C. and KUKSIS, A. (1968). *Lipids*, **3**, 291–300.
70. BRECKENRIDGE, W. C. and KUKSIS, A. (1969). *Lipids*, **4**, 197–204.
71. KUKSIS, A. and BRECKENRIDGE, W. C. (1968). *Dairy Lipids and Lipid Metabolism*. Ed. M. F. Brink and D. Kritchevsky. AVI Publishing Co., Westport, 28–98.
72. MARAI, L., BRECKENRIDGE, W. C. and KUKSIS, A. (1969). *Lipids*, **4**, 562–70.
73. CHRISTIE, W. W. and CLAPPERTON, J. L. (1982). *J. Soc. Dairy Technol.*, **35**, 22–4.
74. PARODI, P. W. (1982). *Lipids*, **17**, 437–42.
75. GRIGOR, M. R. (1980). *Comp. Biochem. Physiol.*, **65B**, 427–30.
76. MORRISON, W. R., JACK, E. L. and SMITH, L. M. (1965). *J. Am. Oil Chem. Soc.*, **42**, 1142–7.
77. MORRISON, W. R. and SMITH, L. M. (1967). *Lipids*, **2**, 178–82.
78. KUKSIS, A., MARAI, L. and MYHER, J. J. (1973). *J. Am. Oil Chem. Soc.*, **50**, 193–201.
79. SCHOMBURG, G., DIELMANN, R., HUSMANN, H. and WEEKE, F. (1976). *J. Chromatogr.*, **122**, 55–72.

80. GROB, K., NEUKOM, H. P. and BATTAGLIA, R. (1980). *J. Am. Oil Chem. Soc.*, **57**, 282–6.
81. TRAITLER, H. and PREVOT, A. (1981). *J. High Res. Chrom. Commun.*, **4**, 109–14.
82. WAKEMAN, S. G. and FREW, N. M. (1982). *Lipids*, **17**, 831–43.
83. MUUSE, B. G. and VAN DER KAMP, H. J. (1984). *Challenges to Contemporary Dairy Analytical Techniques.* Royal Society of Chemistry, London, 192–200.
84. MURATA, T. and TAKAHASHI, S. (1973). *Analyt. Chem.*, **45**, 1816–23.
85. GROB, K. (1979). *J. Chromatogr.*, **178**, 387–92.
86. GROB, K. (1981). *J. Chromatogr.*, **205**, 289–96.
87. HAMMOND, E. W. (1981). *J. Chromatogr.*, **203**, 397–403.
88. HITCHCOCK, C. H. S. and HAMMOND, E. W. (1980). *Developments in Food Analysis Techniques, Vol. 2.* Ed. R. D. King. Elsevier Applied Science, London, 185–224.
89. HAMMOND, E. W. (1982). *HPLC in Food Analysis.* Ed. R. Macrae. Academic Press, London, 167–85.
90. AITZETMULLER, K. (1982). *Prog. Lipid Res.*, **21**, 171–93.
91. KUKSIS, A., MARAI, L. and MYHER, J. J. (1983). *J. Chromatogr.*, **273**, 43–66.
92. MARAI, L., MYHER, J. J. and KUKSIS, A. (1983). *Can. J. Biochem. Cell Biol.*, **61**, 840–9.
93. MYHER, J. J., KUKSIS, A., MARAI, L. and MANGANARO, F. (1983). *J. Chromatogr.*, **283**, 289–301.
94. KUKSIS, A., MYHER, J. J. and MARAI, L. (1984). *J. Am. Oil Chem. Soc.*, **61**, 1582–9.
95. PARRIS, N. A. (1978). *J. Chromatogr.*, **149**, 615–24.
96. PARRIS, N. A. (1978). *J. Chromatogr.*, **157**, 161–70.
97. ATKIN, D. S., HAMILTON, R. J., MITCHELL, S. F. and SEWELL, P. A. (1982). *Chromatographia*, **15**, 97–100.
98. PHILLIPS, F. C., ERDAHL, W. L., NADENICEK, J. D., NUTTER, L. J. and SCHMIT, J. A. (1984). *Lipids*, **19**, 142–50.
99. MACRAE, R., TRUGO, L. C. and DICK, J. (1982). *Chromatographia*, **15**, 476–8.
100. STOLYHWO, A., COLIN, H. and GUIOCHON, G. (1983). *J. Chromatogr.*, **265**, 1–18.
101. STOLYHWO, A., COLIN, H., MARTIN, M. and GUIOCHON, G. (1984). *J. Chromatogr.*, **288**, 253–75.
102. ROBINSON, J. L. and MACRAE, R. (1984). *J. Chromatogr.*, **303**, 386–90.
103. LINDQVIST, B., SJOGREN, I. and NORDIN, R. (1974). *J. Lipid Res.*, **15**, 65–73.
104. SMITH, E. C., JONES, A. D. and HAMMOND, E. W. (1980). *J. Chromatogr.*, **188**, 205–12.
105. NUTTER, L. J. and PRIVETT, O. S. (1967). *J. Dairy Sci.*, **50**, 1194–9.
106. MORRISON, I. M. and HAWKE, J. C. (1977). *Lipids*, **12**, 994–1004.
107. PARODI, P. W. (1980). *Austral. J. Dairy Technol.*, **35**, 17–22.
108. PARODI, P. W. (1982). *J. Dairy Res.*, **49**, 73–80.
109. SHEHATA, A. A. Y., DEMAN, J. M. and ALEXANDER, J. C. (1971). *Can. Inst. Food Technol. J.*, **4**, 61–7.
110. SHEHATA, A. A. Y., DEMAN, J. M. and ALEXANDER, J. C. (1972). *Can. Inst. Food Technol. J.*, **5**, 13–21.

111. TAYLOR, M. W. and HAWKE, J. C. (1975). *N.Z.J. Dairy Sci. Technol.*, **10**, 40–8.
112. MORRISON. I. M. and HAWKE, J. C. (1979). *Lipids*, **14**, 391–4.
113. PARODI, P. W. (1979). *J. Dairy Res.*, **46**, 633–9.
114. PARODI, P. W. (1981). *J. Dairy Res.*, **48**, 131–8.
115. TIMMS, R. E. (1980). *Austral. J. Dairy Technol.*, **35**, 47–53.

6

Wheat Grain Lipids and their Role in the Bread-making Process

P. J. Barnes

RHM Research and Engineering Ltd, High Wycombe, UK

6.1 INTRODUCTION

Lipids represent only 2–3% by weight of wheat grain but play an important role in baking. Although flour lipids have been studied by cereal scientists for many years there is renewed interest owing to several factors. These include a greater interest in the physical aspects such as surface activity, a greater knowledge of lipid-mediated protein aggregation and an increasing need to understand the factors governing the quality of brown flours. Much has been written about the lipid composition of white flour and the possible mechanisms underlying the role of lipids in baking. The purpose of this chapter is to take a broad view starting with the structure and lipid composition of the wheat grain, explaining how these are modified during milling to give the typical composition of flour and finally attempting to provide a concise summary of our understanding of the role of lipids in baking. In this latter part emphasis will be given to physical mechanisms and to recent work.

To conform with present biochemical nomenclature for lipids the following names have been used: triacylglycerol, triglyceride; acylsterol, steryl ester; acylglycosylsterol, esterified sterol glycoside; diacyldigalactosylglycerol, digalactosyldiglyceride; fatty acid, free fatty acid.

6.2 STRUCTURE OF THE WHEAT GRAIN

The wheat grain is usually considered to consist of the endosperm, bran and germ (Fig. 6.1) although this is a simplification as described below.

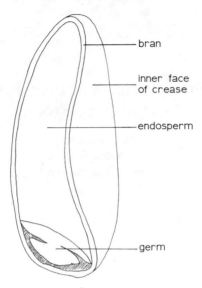

FIG. 6.1. A simplified diagram of a median longitudinal section of the wheat grain.

The endosperm is the inner part of the grain and is the location of almost all of the starch and of those proteins that form gluten. Around this endosperm is a protective layer of fibrous tissues known commonly as bran and towards one end of the grain is found the germ. This terminology is simple and convenient but it is botanically incorrect in certain respects and requires clarification in order to understand what follows.

More accurately, the inner part of the grain is the starchy endosperm and in the wheat grain it is enveloped by a layer one cell thick known as the aleurone layer. This layer differentiates from the rest of the endosperm during grain development and, although it is considered botanically to be a part of the endosperm, it has a very different composition with no significant amount of starch granules or gluten proteins. Lying outside the aleurone layer is the seed coat, or testa, which is a thin, compressed tissue consisting of a pigmented layer sandwiched between two cutinised layers (Fig. 6.2). Further protection is afforded by the pericarp, a fibrous tissue comprising several layers of dead cells. During the milling of white flour the starchy endosperm is recovered to become the flour whilst the aleurone, testa and pericarp largely remain together giving the flakes known to the miller as bran. The part of the grain known as the germ (Fig. 6.3) is comprised primarily of two tissues, the

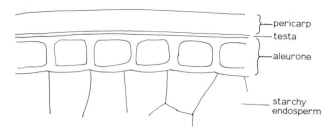

FIG. 6.2. Vertical section through the outer layers of the wheat grain.

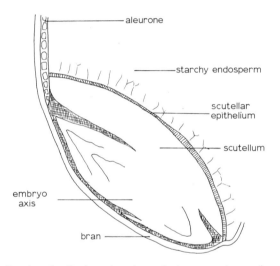

FIG. 6.3. Median longitudinal section through the germ tissue of a wheat grain.

embryo axis and the scutellum, which distribute differently during milling; the embryo axis becomes the miller's germ whereas the scutellum fragments are spread among the other mill fractions.

The starchy endosperm is a storage tissue whose starch, upon germination, would be hydrolysed to sugars by enzymes released from the aleurone and germ, and the sugars transported through the scutellum to provide food for the growing embryo axis. These functions are reflected in the structure and composition of the tissues, the germ and aleurone consisting of living cells rich in enzymes and other proteins, lipids, vitamins and minerals whilst the other tissues are non-viable and comprised mainly of either storage reserves or cell wall polymers.

6.3 COMPOSITION OF LIPIDS IN WHEAT GRAINS

6.3.1 Lipid Classes Present in Wheat Grains

Most of the lipid fraction of the wheat grain consists of acyl esters which are primarily acylglycerols with small quantities of acylated sterols, carotenoids and tocopherols. For detailed reviews of these lipid classes in wheat the reader is referred to the reviews by Morrison.[1,2]

Triacylglycerols are located in oil bodies and represent the major component in lipid extracts from whole wheat grain (Table 6.1). In contrast, the next most abundant lipid class is lysophosphatidylcholine which is found in the starch granules of the endosperm. A number of glycolipids are found in the grain and the major component of these is diacyldigalactosylglycerol believed to be derived from amyloplast membranes, surrounding starch granules during their development. Other than lysophosphatidylcholine, the glycolipids and phospholipids are present mainly as diacyl derivatives but small quantities of the other monoacyl derivatives are found. Monoacylmonogalactosylglycerol is also found as the 6-0-acyl lipid and both phosphatidylethanolamine and lysophosphatidylethanolamine occur as N-acyl esters. Only low concentrations of unesterified fatty acids are found in the intact grain.

Sterols are present as unesterified sterol, acyl sterol, glucosyl sterol and 6-0-acyl glucosylsterol. As in most cereal grains sitosterol is the major sterol and campesterol is the second most abundant. The main carotenoid in the wheat grain is the xanthophyll, lutein, and is found mainly as the monoacyl and diacyl esters. Tocopherols and tocotrienols are mostly unesterified in the wheat grain and consist of α- and β-tocopherol and α- and β-tocotrienol. A number of other minor lipids including ceramides, glycosylceramides, glycophosphoceramides, hydrocarbons and triterpenols have been detected in wheat grain but will not be discussed here. The lipids that are of importance in the baking quality of wheat flour are the acylglycerols, fatty acids, glycolipids and phospholipids; the yellow carotenoids can have an important influence on the colour of flour, bread and pasta products.

6.3.2 Distribution of Lipids among the Grain Tissues

In view of the differences in structure and function of the individual grain tissues it is not surprising to find that the lipid composition is not uniform throughout the grain. Analysis of the lipids in dissected tissues offers far more useful information than does analysis of the whole grain. The following information on lipid distribution is based on the detailed studies carried out by Hargin and Morrison.[3]

TABLE 6.1
CONCENTRATION OF LIPIDS IN WHEAT GRAIN AND IN THE DISSECTED FRACTIONS OF WHEAT GRAIN (mg/g dry weight)[a]

Lipid class	Whole grain	Germ	Aleurone	Starchy endosperm	
				Non-starch lipids	Starch lipids
Acylsterol	0·6	9·3	1·1	0·5	<0·1
Triacylglycerol	8·7	232·0	57·6	1·4	<0·1
Diacylglycerol	0·4	5·0	2·9	0·5	<0·1
Fatty acid	2·6	5·4	2·5	0·5	0·3
Monoacylglycerol	0·7	2·1	0·7	0·8	<0·1
6-0-acylglycosylsterol	0·3		1·6	0·1	<0·1
Diacylmonogalactosylglycerol	0·8	0·8[b]	0·5[b]	0·8	<0·1
Monoacylmonogalactosylglycerol	0·3	0·9[b]	0·6[b]	0·2	<0·1
Diacyldigalactosylglycerol	2·8	1·1[b]	2·1[b]	2·1	
Monoacyldigalactosylglycerol		3·5[b]	2·1[b]		<0·1
N-acyl phosphatidylethanolamine	2·3	0·3		1·4	
N-acyl lysophosphatidylethanolamine	0·8	0·9			
Diphosphatidylglycerol		0·6	0·1		
Phosphatidylglycerol	0·6	2·8	2·3	<0·1	
Phosphatidylethanolamine		4·1			
Phosphatidylcholine	1·9	19·3	8·3	0·5	
Phosphatidylinositol		8·1	2·7		
Lysophosphatidylcholine	4·7	2·6	1·3	0·8	5·0
Other lysophospholipids	1·7	5·0			2·3
Phosphatidic acid	0·1			0·1	0·2

[a] English hard winter wheat, variety Flinor. Calculated from the data of Hargin and Morrison[3] and corrected to the nearest 0·1 mg/g.
[b] Tentative identification.

The starchy endosperm accounts for 40–50% of the total grain lipids whilst the remainder is divided in approximately equal proportions between the aleurone and germ. Taking into account the different relative proportions of germ and aleurone in the grain this means that the germ has a higher concentration of acyl lipids, in the region of 300 mg/g dry weight, than does the aleurone with 100 mg/g (in the following discussion lipid is taken to mean acyl lipid because the non-saponifiable fraction represents only a very minor proportion of the total lipids except in the pericarp). Commercial mill germ has a much lower lipid content than does dissected germ for reasons discussed below. In contrast, pericarp contains only about 13 mg/g and starchy endosperm about 20 mg/g of lipids.

Although germ, aleurone and pericarp have quite different total lipid contents the non-polar fraction represents around 80% of the total in each tissue. Triacylglycerols are the major lipid class in germ and aleurone (Table 6.1) and account for the high content of hexane-extractable oil in mill germ and mill bran. Unesterified fatty acids are present at low concentrations in the intact tissues. There is little information on the pericarp lipids but they appear to consist of a mixture of acylglycerols and fatty acids together with a large proportion of unsaponifiable lipids; these probably derive from the original living cells of the pericarp, present during grain development, combined with the waxy cuticular material on the outermost surface of the grain.

The polar fraction is almost completely phospholipid in the germ but in the aleurone small amounts of glycolipid are also present. About half of the phospholipid in the germ and aleurone is phosphatidylcholine and the glycolipid fraction of the aleurone contains substances tentatively identified as diacyl and monoacylgalactosylglycerols and acylglycosylsterol.

Unlike the other tissues, the total lipids of the starchy endosperm contain a much greater proportion of polar lipids compared to non-polar lipids and this is largely due to the preponderance of one particular lipid, lysophosphatidylcholine (Table 6.1). However, analyses of the total lipids of the starchy endosperm can mask important differences in the distribution of the lipid classes and these differences are important in both biochemical and technological studies of cereal grains. The endosperm lipids of cereals can be conveniently divided into two main fractions: starch lipids and non-starch lipids.[1,4] Starch lipid is the term frequently used to denote the lipid fraction found within the starch granules and which requires disruption of the granule structure by hot alcohol–water mixtures for thorough extraction; they are always monoacyl lipids and are thought to exist as amylose inclusion complexes. In contrast, the oil body and membrane-derived lipids can be rapidly extracted into cold solvents and are designated non-starch lipids. It has been necessary to distinguish between starch internal lipids (those located within the granule and difficult to extract, as described above) and starch surface lipids which are bound to the surface of the granules.[5]

The lysophosphatidylcholine in the wheat endosperm is mostly present as starch internal lipid and is a relatively minor component in the non-starch lipids. Protected within the starch granule, it is not susceptible to enzymatic hydrolysis of the acyl ester bond or to oxidation or chlorination of the linoleyl chain but may have an important influence on the physico-chemical properties of the starch granules. Whereas the starch internal lipids are almost exclusively lysophospholipids, the non-starch lipids comprise non-polar lipids (mainly acylglycerols), glycolipids and phospholipids. Unlike in the other grain tissues, a major proportion of the phospholipids of the non-starch fraction are present as the N-acyl esters of phosphatidylethanolamine and lysophosphatidylethanolamine. The most characteristic components of the non-starch lipids, and probably the most interesting from the technological point of view, are the glycolipids, and, in particular, the acylgalactosylglycerols. In living tissues these lipids are typically found in chloroplast membranes and have been identified in the amyloplast membranes of potato tuber.[6] Both chloroplasts and amyloplasts are cytoplasmic organelles that are capable of synthesising starch although in the chloroplast the major function is photosynthesis and little starch accumulates. In the developing wheat endosperm the amyloplasts are the plastids within which the starch granules are formed and it is probable that the acylgalactosylglycerols in the mature endosperm are present in amyloplast membrane residues. The ability of diacyldigalactosylglycerol in the presence of water to form a lamellar liquid–crystalline phase is not only relevant to its function in the biological membrane but may be an important factor in its effect on the baking quality of flour as discussed below.

6.3.3 Lipid-degrading Enzymes in the Wheat Grain

In any discussion of the lipid composition of cereal grains it is important to take into account the presence of enzymes capable of catalysing the hydrolysis or oxidation of the lipids during storage and processing (and also during analysis if precautions are not taken). The enzymatic degradation of cereal lipids has been reviewed recently by Galliard[7,8] and will not be discussed in detail here.

Several types of lipid hydrolase activity have been found associated with wheat grain. Long-chain triacylglycerols are hydrolysed when milled grain (wholemeal flour), mill germ or bran are stored at moisture contents of above 5% but it is not certain to what extent the microbial flora of the grain might contribute to this activity. This hydrolytic activity appears to be concentrated in the bran.[8,9] The activity is relatively low compared to that of the oat grain, for instance, and should not be confused with the much higher 'esterase' activity found when using water-soluble substrates such as short-chain acyl esters. An intermediate type of hydrolytic activity present in wheat is that which catalyses hydrolysis of monoacylglycerols in the monodisperse state but not triacylglycerols at an oil–water interface. The triacylglycerol hydrolase is important technologically because it is responsible for accumulation of fatty acids in stored wheat products. Phospholipase D is present in wheat and shows both hydrolase and transferase activities, depending upon the availability of water.

Oxidation of polyunsaturated fatty acids can occur in disrupted wheat grain tissues both with and without the intervention of enzymatic catalysis but the presence of lipoxygenase in the tissues is responsible for rapid oxidation of unesterified linoleic and linolenic acids and monoacylglycerols under appropriate conditions of moisture and temperature. Oxidation can lead to a variety of products, some of which affect flavour and aroma of wheat products or may be detrimental to functionality. Wheat lipoxygenase is concentrated in the germ and the potential lipoxygenase activity of wheat products is thus influenced by the proportion of germ that they contain.

6.4 RELATIONSHIP BETWEEN GRAIN LIPID COMPOSITION AND FLOUR LIPID COMPOSITION

In the commercial milling of wheat grain the grains are broken and ground down to a suitable particle size distribution to yield wholemeal flour and this is described as 100% extraction because all the grain is converted into flour with no by-products. However, in many industrialised countries wholemeal flour is not the major product and the bulk of the flour produced is almost pure endosperm ('white flour') at an extraction of 70–80% as a result of fractionation into endosperm, bran and germ. A range of intermediate extraction bread flours ('brown flours') are also produced to satisfy the increasing consumer demand for foods with higher dietary fibre content. In addition to use in bread and other baked goods white flour is finding increasing use in industrial applications, particularly in the production of wheat gluten and starch. Wheat gluten is the protein fraction obtained when the starch granules are washed out from a dough made from flour and water. Commercial wheat gluten is finding increasing use in Europe for the protein fortification of flour milled from European wheat to avoid the need for using imported high-protein North American wheat. Likewise, the wheat starch produced from gluten washing-out plants is replacing imported maize starch in the UK.

It might be expected that the lipid composition of a commercial wholemeal flour would be exactly the same as the wheat from which it was milled because there has been no fractionation of the grain tissues. In fact, there is no information in the literature to show whether or not this is true and it remains to be seen whether any hydrolysis or oxidation of lipids occurs during milling. However, once the grain has been disrupted lipid degradation proceeds (unless the flour is stored at low temperature) and the lipid composition will begin to deviate from that of the intact grain. It is likely that in wholemeal flour there will have been some redistribution of the lipids during milling with transfer from the lipid-rich tissues to the endosperm fragments as is known to occur in white flour (see below).

In the milling of flours of extraction rate less than 100%, bran and germ are selectively removed and the flour must, therefore, differ in lipid composition from the intact grain. As more bran and germ are removed the lipid composition of the flour will approach that of the pure starchy endosperm and it might be expected that in a white flour only endosperm lipids are present. In practice this is not the case and even the highest grades of white flour (patent flours) will contain a proportion of bran and germ lipids. To help understand how this occurs there follows a brief description of the flour milling process and the redistribution of lipids during milling.

If dry wheat is ground to a fine flour it is impossible, using any commercial equipment, to make an efficient separation of endosperm particles from bran and germ. Instead, a sequence of breaking, grinding and separating operations are used (Fig. 6.4). The grains are first broken open to yield endosperm fragments on a series of fluted 'break' rolls and the endosperm is then gradually ground to a fine flour by passage through a number of smooth 'reduction' rolls. Bran contaminating the endosperm is less friable and tends to remain in the form of larger particles so that the released material from each grinding step can be

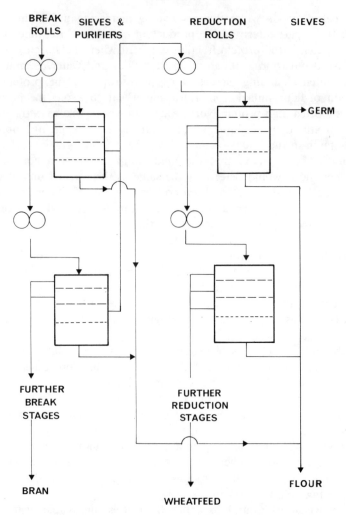

FIG. 6.4. A simplified schematic diagram of the flour milling process.

fractionated on sieves. The finest fraction contains the purest endosperm and represents a patent flour stream; the coarse material passes to the next reduction roll for further grinding and fractionation. Eventually, the coarse fraction is depleted in starchy endosperm and becomes 'wheatfeed' which is a bran-rich product used in animal feeds. Flours are produced at each grinding stage and range from the almost pure endosperm, white high-grade products to the brown lower-grade with relatively high proportions of germ and bran content. Most of the flour is in the high-grade category but some of the lower-grade must be blended in to ensure an economic yield of white flour with a high protein content. The germ is a soft tissue and is compressed into thin flakes on the rolls, allowing a germ fraction to be removed if required and purified as a separate product. Otherwise the germ passes into the wheatfeed to become a component of animal feeds or is retained in the flour in the case of wholemeal flours.

6.4.1 Redistribution of Grain Lipids during Milling

Among the different flour streams from a mill those with the darkest colour, and hence the highest content of bran particles, also have the highest oil content (Fig. 6.5). The obvious explanation is that the bran and germ particles found in the darker flours have a higher lipid content than the endosperm. However, this is not the only contribution to the high oil content. As the fragments of bran, germ and endosperm pass between the mill rolls they are compressed with great force resulting in

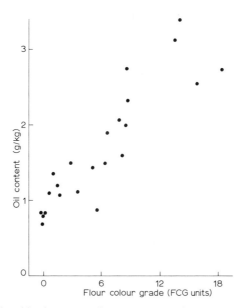

FIG. 6.5. Relationship between oil content and flour colour grade for the individual machine flours from a commercial mill. Oil content is expressed as hexane-extractable acyl lipids. Adapted from Morrison and Barnes.[10]

the expression of lipids from the oil-rich tissues and transfer to the endosperm particles. Thus, the final flour contains (1) the endosperm lipids, (2) germ and bran lipids within tissue fragments and (3) germ and bran lipids transferred to endosperm particles.

The transfer of germ oil to endosperm during milling has long been suspected from indirect evidence. The germ fraction that is collected after compression into flakes and separation by sieving consists of the embryo axis and not the scutellum, the latter being more friable and difficult to separate.[11,12] Whereas the oil content of a dissected embryo axis is about 150 mg/g,[13] that of the purest commercial germ is typically no more than 100 mg/g, suggesting that the oil lost from the germ may have been transferred to the flour. In later work, a comparison was made between flours milled from wheat before and after degerming the grain. It was estimated that approximately 30% of the oil in the normal flour was germ oil and that half of this could be accounted for by transfer from the germ during milling.[14]

6.4.2 Tocopherol Markers for Germ and Aleurone Lipids

Recent studies using specific chemical markers have confirmed the presence of germ lipids and aleurone lipids in white flour.[10] The markers are tocopherols and tocotrienols which are fat-soluble phenolic compounds responsible for the vitamin E content of plant tissues.

Wheat grains contain α-tocopherol (α−T), β-tocopherol (β−T), α-tocotrienol (α-T-3) and β-tocotrienol (β-T-3) but it was known from analysis of milling products that the relative proportions were probably not uniform throughout the grain. Development of a rapid method based on high-performance liquid chromatography combined with fluorescence detection[15] allowed a detailed study to be made of the distribution of these compounds in the grain.[16] It was found that the germ contained a very high concentration of α-T and β-T but the tocotrienols were undetectable (Table 6.2). In contrast, the bran and starchy endosperm both contained α-T-3 and β-T-3 but with only minor amounts of α-T and β-T. It is notable that 98% of the α-T present in the wheat grain was located in the germ, confirming that α-T could be used as a 'marker' for the presence of germ lipids in flour (Fig. 6.6). The β-T-3 was equally distributed between the bran (aleurone) and starchy endosperm whereas the α-T-3 was found predominantly in the aleurone and could be used to gain information on the redistribution of aleurone lipids during milling although it is less specific compared to α-T. These markers are fat-soluble and can be present in flour either as components of germ and aleurone

TABLE 6.2
CONCENTRATION OF TOCOPHEROLS AND TOCOTRIENOLS IN DISSECTED WHEAT GRAIN FRACTIONS (mg/kg dry weight)[a]

Fraction	α-T[b]	β-T	α-T-3	β-T-3
Germ	255·6	114·4	<2	<2
Pericarp, testa and aleurone	0·5	<0·4	10·0	68·6
Starchy endosperm	0·07	0·1	0·5	13·5

[a] From Morrison et al.[16]
[b] α-T: α-tocopherol; β-T: β-tocopherol; α-T-3: α-tocotrienol; β-T-3: β-tocotrienol.

tissue fragments or as components of oil transferred from the tissues to the endosperm during milling as described above. Their use is primarily as markers for the germ and aleurone lipids but they can also be used in conjunction with other analyses to gain information on the distribution of the tissues.

Tocopherol analysis has been used to examine the 23 flour streams from a commercial mill.[16] The α-T values ranged from 2·2 to 42·3 mg/kg and were closely correlated with the oil content showing the extent to which germ lipid influences the lipid content of flour streams (Fig. 6.7). Using the α-T data it was possible to calculate, firstly, the composition of a hypothetical flour in the absence of lipids from germ and aleurone and, secondly, the composition of the lipid derived from the germ and aleurone. Most of the variation in lipid composition between the flour streams was accounted for by triacylglycerols, diacylphospholipids and tocopherols whereas very little variation was shown by the glycolipids and the N-acyl phospholipids (Fig. 6.8).

The results are in accord with the measured lipid composition of dissected endosperm, germ and aleurone. The calculated composition of the flour in the absence of germ and aleurone lipids is close to that obtained for dissected endosperm (Table 6.3). Similarly, the composition calculated for the lipid fraction derived from germ and aleurone approximates to that of germ oil and aleurone lipids. The content of glycolipids and N-acyl phospholipids in flour is governed mainly by the levels present in the starchy endosperm before milling. In contrast, the amounts of acylsterols, triacylglycerols, diacylglycerols, fatty acids, diacyl phospholipids and α- and β-tocopherols are dependent to a certain extent on the proportion of germ and aleurone lipids in the flour. As described below, the ratio of non-polar to polar lipids and the composition of

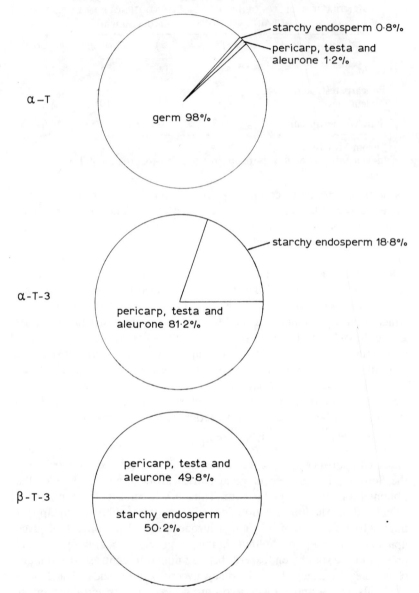

FIG. 6.6. Distribution of tocopherols among the major tissues of the wheat grain.

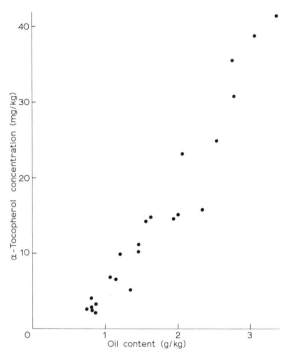

FIG. 6.7. Relationship between α-T concentration and oil content (hexane-extractable acyl lipids) for the individual machine flours from a commercial mill. Adapted from Morrison and Barnes.[10]

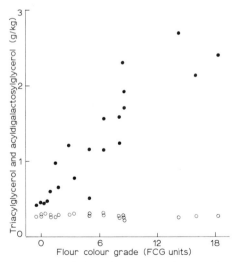

FIG. 6.8. Relationship of triacylglycerol content (●) and acyldigalactosylglycerol content (○) to the flour colour grade of the individual machine flours from a commercial mill. Adapted from Morrison and Barnes.[10]

TABLE 6.3
COMPARISON OF ENDOSPERM NON-STARCH LIPID COMPOSITION CALCULATED FROM MILLSTREAM DATA WITH THE COMPOSITION OBTAINED BY ANALYSIS OF DISSECTED TISSUE[a]

Lipid class	Proportion of total endosperm non-starch lipids (%)	
	Calculated	Determined
Acylsterol	3	2
Triacylglycerol	26	28
Diacylglycerol and fatty acid	9	14
Monoacylglycerol and acylglycosylsterol	6	9
Monogalactosylglycerols	8	6
Digalactosylglycerols	24	24
N-acyl phosphatidylethanolamine	8	9
N-acyl lysophosphatidylethanolamine	4	6
Diacyl phospholipids	6	4
Monoacyl phospholipids	5	

[a] Adapted from Tables 7.2 and 7.6 in Morrison and Barnes;[10] original data recalculated to give percentage composition.

glycolipids in flour influence baking quality. It may be possible to alter the lipid composition of the endosperm through plant breeding and Morrison and coworkers[17] have recently begun to unravel the control of petroleum-extractable glycolipid composition by the group 5 chromosomes of wheat. The content of non-polar lipids in flour is partly dependent on the concentration in the endosperm but will also be influenced by factors affecting the redistribution of germ and aleurone tissue and lipids during milling, such as grain moisture content and intensity of milling.

The moisture content of the wheat grain used in milling is carefully controlled because an optimum moisture level is needed in order to obtain the highest yield of white flour with the lowest bran content. However, moisture content also affects the redistribution of germ tissue. At low moisture the germ is more easily pulverised, separates less efficiently from the endosperm fractions and eventually reaches the white flour streams from the reduction rolls. By using a higher moisture content the germ can be kept to a larger particle size and prevented from contaminating the flour[10-12] (Table 6.4). In addition to increasing the content of non-polar lipids in the flour, germ affects baking quality directly.[18]

TABLE 6.4
THE EFFECT OF WHEAT MOISTURE CONTENT ON THE DISTRIBUTION OF α-T INTO MILLSTREAMS ON A BÜHLER EXPERIMENTAL MILL

Millstream		α-T (mg) from 10 kg of grain	
		11·8% moisture	18·5% moisture
1st break		5	1
2nd break		8	4
3rd break		2	6
	Total	15	11
1st reduction		14	9
2nd reduction		23	8
3rd reduction		22	8
	Total	59	25
Bran		9	84
Wheatfeed		75	29
	Total	84	113

6.5 STORAGE STABILITY OF WHEAT FLOUR

The involvement of lipid hydrolysis and oxidation in the storage deterioration of foodstuffs is well known[19] and it is not surprising that studies of the storage stability of bread flour have focussed on the role of lipids.

Detailed studies have been made of the change in baking quality and lipid composition of white flours during prolonged storage.[20] The effect of storage on loaf volume depended on the type of flour, the baking process and the presence or absence of added fat; in some cases the deterioration in quality was preceded by an increase in loaf volume in the early stages of storage. The fatty acid content of the flours increased on storage and there were changes in the composition of the lipids as would be expected if lipolysis were taking place. Notable among these was the drop in concentration of diacyldigalactosylglycerol. There was a decrease in the total linoleic acid and linoleyl groups in the flour and an increase in mono- and dihydroxyoctadecadienoic acids suggesting that oxidation took place. Storage of flour also brought about a change in the proportions of 'free' and 'bound' lipids when the flour was made into dough and it was postulated that the high concentrations of fatty acids influence lipid–protein associations during dough-mixing.[21]

Although fatty acids accumulate in stored flour and fatty acids added to flour adversely affect baking quality there is almost no direct evidence that changes in the lipid fraction during storage are a cause of changes in baking quality. Cuendet and coworkers[22] showed that the lipid fraction was partly responsible for the deterioration in baking quality of white and wholemeal flours stored at 38°C. Recent work has confirmed that both the extractable lipid fraction and the unextractable fraction of wholemeal flour stored at 20°C are involved in the loss of baking quality during storage.[23] Wholemeal flour is more susceptible to storage deterioration than is white flour and current research is helping to identify the flour fractions most responsible for rancidity and loss of baking quality in brown flours.[8,24,25]

The results of this research show that lipolysis occurs in the bran fraction of the wholemeal flour during storage, with the accumulation of unesterified fatty acids. During dough-mixing, the unesterified polyunsaturated fatty acids are oxidised by the germ lipoxygenase with formation of conjugated fatty acid oxidation products. This oxidation competes for molecular oxygen with other reactions occurring in the dough and thus influences the redox state of the dough. Furthermore, it is possible that polyunsaturated fatty acids released during storage undergo a slow, but significant, autoxidation and that the resulting peroxides, or their products, react with the endosperm proteins. This modification of the proteins might then be manifest as a change in gluten functionality during dough-mixing.

6.6 THE ROLE OF FLOUR LIPIDS IN BAKING OF BREAD

6.6.1 The Bread-making Process

In making bread the flour and other ingredients are mixed with water to form the coherent viscoelastic mass known as dough. Microscopic air bubbles are incorporated during mixing and the endosperm proteins develop into a continuous gluten matrix. In commercial baking the dough is compressed by passage between rollers, which sub-divides the air bubbles and ensures a more uniform size distribution, then allowed to stand whilst the bubbles grow in size by accumulation of carbon dioxide from yeast activity and the dough increases in volume. Finally, the dough is heated causing the bubbles to expand further due to gas expansion, water vapour and accelerated yeast activity, the dough rises and the structure sets as a result of gelatinisation of starch granules. There are several different bread-baking processes but they all embody these fundamental principles and it will not be necessary to complicate the following discussion by reference to different commercial methods.

An essential requirement in baking is that dough-mixing produces a foam that (a) remains stable during bubble expansion and heating until the open-celled structure sets, and (b) is strong enough to support the starch and, in the case of brown flours, the bran. Lipids may influence the properties of dough by modulating the interfacial characteristics, such as wetting of the flour components during mixing, stabilisation of the gas cell membrane and formation of interfaces between protein lamellae.[26, 27] This influence on interfacial properties can explain why lipids have important effects on baking quality although present in the flour at levels of only 1–2%. It also can account for the effect of added emulsifiers at even lower concentrations.[28]

The following account makes no attempt to describe all the research that has been done to unravel the mechanisms by which lipids influence baking quality. Comprehensive reviews have been published elsewhere.[29–36] Instead it is intended to introduce some of the more recent concepts and to emphasise the importance of the physical properties of flour lipids. For further information the reader should consult the earlier reviews.

6.6.2 Observed Effects of Flour Lipids in Bread-making

It is now well substantiated that the polar lipids of flour have a beneficial effect during baking and the non-polar lipids are detrimental.[29, 34, 35] Glycolipids are more effective than phospholipids and diacyldigalactosylglycerol is particularly effective. Among the non-polar lipids, fatty acids have the greatest detrimental effect and linoleic acid has more effect than palmitic acid. The starch internal lipid (lysophosphatidylcholine) is localised within the granules and is thought not to influence dough-mixing or the early stages of baking.

Flour from which the bulk of the non-starch lipids have been removed ('defatted flour') yields bread with good loaf volume and fine, uniform crumb texture. Loaf volume decreases with increasing amounts of the extracted non-polar lipids added back to the flour. Small amounts of the extracted polar lipids added back to the defatted flour cause a sharp decrease in loaf volume but further addition increases volume up to or beyond that of the original unextracted flour (Fig. 6.9); a similar relationship exists between loaf volume and polar lipid content of the flour when shortening fat is used in the baking test. Addition of the unfractionated lipid extract back to the defatted flour results in a volume–lipid content curve similar to that for polar lipids alone but with the minimum loaf volume shifted to higher lipid content. Crumb texture approximately follows the shape shown by the loaf volume curve. The free (petroleum-extractable) polar lipids of flour have a greater beneficial effect on loaf volume than do the bound polar lipids and this is ascribed to the greater proportion of glycolipids in the former. Addition of glycolipids improves loaf volume in the presence or absence of shortening but phospholipids only improve volume in the presence and have no effect in the absence of shortening.

When flour lipids are added to unextracted flour the effect on loaf volume is much less than for defatted flour except in the case of diacyldigalactosylglycerol and lysophosphatidylcholine which bring substantial improvement. With unextracted flour and shortening present the

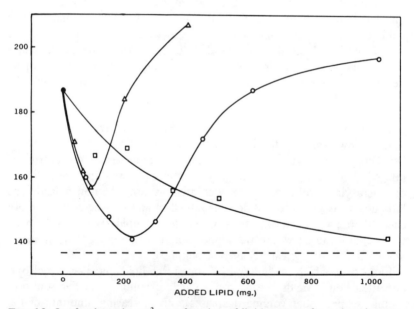

FIG. 6.9. Loaf volume in cm^3 as a function of lipid content, for polar plus non-polar lipid (O), polar lipid (△) and non-polar lipid (□). Dashed line represents volume at end of proof. Additions were made to a dry flour weight of 30·2 g.[37]

non-polar lipids show a slight detrimental effect but the polar lipids increase loaf volume.

It is notable that, for a flour with satisfactory quality and quantity of protein for baking, an increase in the content of polar lipids in the flour can increase loaf volume to a much greater extent than can an increase in protein on a weight for weight basis.

6.6.3 Correlations between Baking Quality and the Lipid Composition of Wheat Grain and Flour

One way of elucidating how flour lipids are involved in baking quality has been to alter the lipid composition of flour by extraction and reconstitution as described above. A second way is to determine whether or not any correlation exists between baking performance and the 'natural' lipid composition of flour.

No significant correlation was found between loaf volume and either the total petroleum-extractable lipids (free lipids) or the total non-starch lipids.[29] However, recent studies have shown close relationships of loaf volume to the content of polar lipids in the petroleum-extractable lipid fraction.[38,39,40] Using either white or whole grain flours milled from North American winter wheats and Canadian spring wheats, loaf volume shows a highly significant positive correlation with polar lipids and galactolipids and negative correlation with the ratio of non-polar lipids to polar lipids and non-polar lipids to galactolipids (Table 6.5). Strong correlations are observed after correcting the data to take account of protein content and they are not due merely to a fortuitous correlation between protein content and lipid content. However, no significant correlation is found between any lipid parameter and loaf volume for UK wheats.[41]

As explained in Section 6.4.2, the glycolipid content of flour is controlled principally by the amount present in the endosperm of the intact grain whereas the content of triacylglycerols and diacylphospholipids is also influenced by the amount of germ and aleurone lipids in the flour. Thus, it may be possible to alter the amounts of petroleum-extractable galactolipids in flour through programmes of wheat breeding.[17]

6.6.4 Mechanisms by which Lipids Influence Baking Quality

It is generally accepted that flour lipids influence baking quality primarily through their interaction with the other flour components and dough ingredients. A number of theories have been proposed to explain the effects of lipids during dough-mixing but none of these is fully proven

TABLE 6.5
CORRELATION COEFFICIENTS FOR LOAF VOLUME WITH CONCENTRATION OF LIPID FRACTIONS IN THE FLOUR[a]

Hexane-extractable lipids	0·556 **
Hexane-extractable non-polar lipids (NL)	0·455 **
Hexane-extractable polar lipids (PL)	0·871 **
Hexane-extractable glycolipids (GL)	0·874 **
Hexane-extractable phospholipids	0·758 **
Bound lipids[b]	0·286 *
Total lipids[c]	0·558 **
Ratio NL/PL[d]	−0·976 **
Ratio NL/GL[d]	−0·968 **

[a] From Békés et al.[40] Calculated from data for flours milled from 77 Canadian spring wheats (*, ** significant at the 5% and 1% levels of probability, respectively).
[b] Fraction extracted into cold water-saturated butan-1-ol after extraction of the flour with hexane.
[c] Sum of bound lipids and hexane-extractable lipids.
[d] After correction of loaf volume for protein content.

even after about 60 years of research. Many of the studies were based on gluten washed out from dough rather than the dough itself and all involved some form of fractionation which might have altered the relationship between lipids and other dough components. These studies have usefully increased our understanding of the role of lipids in baking but have been limited by the lack of suitable non-destructive techniques to examine the interactions between lipids and other components in the intact dough.

The theories concerning the role of lipids in the baking process can be divided into three main areas; these are (1) interfacial activity of the lipids, (2) formation of liquid–crystalline phases and (3) association with proteins.

Interfacial activity

In order to bring about a stable dispersion of one phase in another it is necessary to reduce the interfacial tension between the two phases. This is achieved by the presence of surfactant molecules which concentrate at the interface and exert a spreading pressure in opposition to the interfacial tension. Typical examples of this phenomenon are the stabilisation of gas–water foams, oil–water emulsions and solid–liquid suspensions. Bread dough is a combination of these dispersed systems and it is to be expected that molecules with a high interfacial activity could have a profound influence on baking performance.[27,28,32]

In particular, the foamed structure that dough represents must remain stable during gas cell expansion and heating, survive moulding operations, be strong enough to 'carry' the starch fraction and must allow transfer of carbon dioxide from the liquid to the gas phase. The importance of interfacial activity is demonstrated by the dramatic effect of commercial emulsifiers on loaf volume.

The dispersed gas phase in the dough is finely-divided and surrounded by aqueous phase within which the soluble components and the yeast are concentrated.[27] The gas and liquid are bounded by an interface at which surfactant molecules will form a monomolecular layer and reduce the surface tension. As the gas cells expand the surface area of interface increases and more surfactant molecules move from the bulk aqueous phase to the interface and maintain the low surface tension. If no more surfactant is available the surface tension will rise causing increased pressure in the cell and favouring coalescence of the cells, loss of the fine cellular texture and escape of gas.

Proteins can lower interfacial tension and may be the surfactants responsible for stabilisation of the fine cell structure in doughs made from chloroform-extracted flour.[42] However, in unextracted flour the polar lipids, which have greater spreading pressures, displace proteins from the interface and the foam is lipid-stabilised. At intermediate levels of polar lipid content mixed interfacial films of lipid and protein are formed but these are unstable and baking performance is poor. This can explain the observation, described in Section 6.6.2, that greatest loaf volume is achieved with normal flour and flour from which the polar lipids have been extracted but loaf volume passes through a minimum at an intermediate content of polar lipid in partly reconstituted flours.

The galactolipids and phospholipids in flour possess the necessary amphiphilic properties for generating high spreading pressures at interfaces whereas the triacylglycerols do not. This is in accord with the experimental observations that flour polar lipids have beneficial effects in baking whereas the flour non-polar lipids do not.

In addition to the gas–liquid surface other important interfaces exist in dough. The aqueous phase surrounding the dispersed gas also interfaces with the dispersed soild phase and surfactant molecules will adsorb in the interfacial film where this is in contact with hydrophobic surfaces such as non-polar regions of proteins. Also, the flour triacylglycerols and added shortening will be emulsified by the polar flour lipids or by interaction with proteins.

Liquid–crystalline phases

Triacylglycerols alone do not form lipid–water phases but the polar lipids can form geometrically ordered phases with water. Such phases exhibit properties of both liquids and crystals and the exact liquid–crystalline structure depends on the type of lipid, the water content and the temperature.

Isolated lipid fractions from wheat flour and gluten interact with water to form multiphase mixtures of liquid–crystals, oil and water, the proportions depending on the relative amounts of lipid and water (Fig. 6.10).[26, 32] The strong influence of lipid structure is shown by the behaviour of diacylmonogalactosyl- and diacyldigalactosylglycerols; the former produce a reverse-hexagonal phase and the latter a lamellar phase (Fig. 6.11). With further increase in water content the monogalactolipid

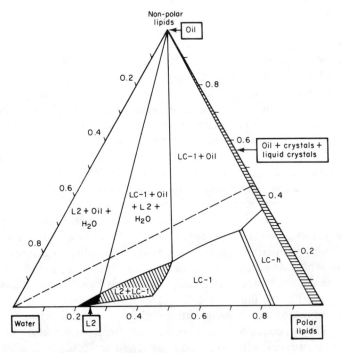

FIG. 6.10. Phase diagram of the interaction between isolated wheat flour lipids and water at 20°C. LC-1, lamellar liquid-crystal phase; LC-h, hexagonal liquid-crystal phase; L2, reverse micellar phase of water dispersed in a continuous lipid phase (from Carlson et al.).[43]

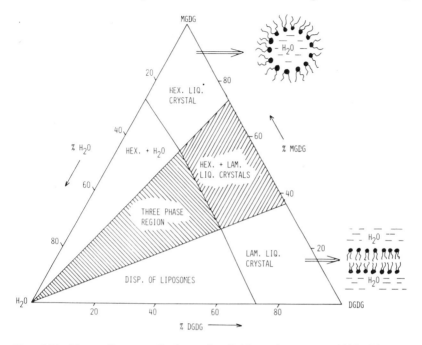

FIG. 6.11. Phase diagram of wheat glycolipids and water at 20°C. MGDG, diacylmonogalactosylglycerol; DGDG, diacyldigalactosylglycerol (from Larsson and Puang-Ngern).[44]

simply forms a separate water phase in addition to the liquid–crystalline phase whereas the digalactolipid undergoes a transition from the lamellar phase to a dispersion of liposomes.

These observations have been made using lipids extracted from flour but it is quite possible that liquid–crystalline phases are formed in the aqueous phase of wheat flour dough in the way that some commercial emulsifiers form liquid crystals with the water-phase of foodstuffs.[28] The water content at which the transitions from one structure to another occur may be different to those of model systems owing to the presence of other components dissolved in the dough water.

The particular relevance of liquid–crystalline phases to bread-making performance is that the lamellar phase is much more effective than many other forms of lipid at providing a reservoir of surfactant molecules for adsorption at a gas–liquid interface.[32] During expansion of the gas cells in dough the interface area increases and a proportional increase in surfactant molecules must occur at the interfacial film if the surface tension is to be kept low. The lamellar liquid–crystalline phase could satisfy this requirement because it forms a liposomal dispersion at high water content or as a result of increase in temperature and the dispersion can continuously yield a liquid-condensed monolayer at an expanding air–water interface. Diacyldigalactosylglycerol, the most abundant polar lipid in wheat flour, is the most effective of the flour lipids tested at increasing loaf volume and the concentration of galactolipids in petroleum extracts of flour has been found, in at least one study,[39] to be closely correlated with baking quality. It is possible that these properties are related to the ability of this lipid to (a) readily form lamellar liquid–crystals and liposomal dispersions with water and (b) to form condensed monolayers with high spreading pressures at interfaces.

It has been suggested from X-ray diffraction data that lipids in the gluten matrix are present as lamellar liquid–crystalline phases,[26] but this information relates to gluten washed out from dough and, therefore, it is not yet certain whether such a relationship between protein and lipid exists in the dough. However, the ability of liquid–crystalline phases to solubilise triacylglycerols could provide a mechanism for the incorporation of fats into the hydrated gluten.

Association of lipids with proteins

Considerable attention has been given to the decrease in extractability of lipids that occurs on mixing flour into dough.[45] The fraction that is extractable from the freeze-dried dough with petroleum solvents is known as 'free lipid' and the fraction that is subsequently extracted by cold polar solvents is known as 'bound lipid'. The proportion of bound lipid as a percentage of the total lipid increases when flour is wetted and mixed into a dough. When gluten is washed from the dough it is found to contain a large proportion of the flour non-starch lipid and most of this lipid is not extractable in hexane. It has long been assumed that the decrease in extractability is due to the lipid molecules associating with hydrophobic regions of the proteins. This association would be expected to occur when water is added to flour but cannot alone explain reduced extractability of triacylglycerols from freeze-dried dough or gluten. The association would be disrupted by hexane and the lipid would be extracted unless it was mechanically occluded or penetration of the solvent was otherwise prevented.

The direct cause of reduced extractability of flour lipids from freeze-dried dough is not known and the significance of lipid-binding with respect to baking performance is not clear. However, certain low molecular weight proteins, when isolated from freeze-dried dough or gluten, are found to be associated with lipids and the association persists through several fractionation procedures.[31,39] In addition, the presence of the lipid promotes and maintains aggregation of certain proteins into high molecular weight complexes. A protein that associates with triacylglycerol has been isolated from dough and proteins associating with polar and non-polar lipids have been isolated from gluten.[46-51]

There is some evidence which suggests that lipid-binding in dough has an adverse effect on baking performance. Dough-mixing at very high work-input under a nitrogen atmosphere results in high levels of bound lipid and low loaf volumes whereas in air the lipid binding is much less and loaf volume higher.[52] Storage-deterioration of flour and addition of fatty acids to flour decrease loaf volume and increase lipid binding in the dough.[20] However, it is possible that these relationships are merely fortuitous and do not represent a cause and effect relationship. It has been suggested recently that the association between digalactolipids and proteins and the resulting aggregation of proteins may have an important role in baking performance, and that this might explain the beneficial effect of diacyldigalactosylglycerol.[48] In the absence of direct evidence for this hypothesis, the alternative view could be held that the association of galactolipids with protein could render these important lipids unavailable for participation in formation of stable gas–liquid interfaces and the association, or binding, would then have an adverse effect on baking performance. More information is needed on the relationship between lipid-binding and baking quality and on the interaction of lipids and proteins in unfractionated wet dough.

It is evident that the role of wheat flour lipids in baking is far from clear. There is no doubt that the lipids do have important effects but more research is needed to elucidate the mechanism of the effects. Until very recently, almost all research on this topic was confined to low extraction white flours. However, changes in the types of bread consumed have influenced the future direction of research and more attention must now be given to brown flours.

6.7 ACKNOWLEDGEMENTS

The author wishes to thank Dr T. Galliard for discussions and Dr J. Edelman, Director of Research, for permission to publish this review.

6.8 REFERENCES

1. MORRISON, W. R. (1978). *Adv. Cereal Sci. Technol.*, **2**, 221–348.
2. MORRISON, W. R. (1983). In *Lipids in Cereal Technology*. Ed. P. J. Barnes. Academic Press, London, 11–32.
3. HARGIN, K. D. and MORRISON, W. R. (1980). *J. Sci. Food Agric.*, **31**, 877–88.
4. ACKER, L. (1974). *Getreide Mehl Brot*, **28**, 181–7.
5. MORRISON, W. R. (1981). *Starke*, **33**, 408–10.
6. FISHWICK, M. J. and WRIGHT, A. J. (1980). *Phytochemistry*, **19**, 55–9.
7. GALLIARD, T. (1983). In *Lipids in Cereal Technology*. Ed. P. J. Barnes. Academic Press, London, 111–47.
8. GALLIARD, T. (1986). In *The Chemistry and Physics of Baking*. Ed. J. M. V. Blanshard, P. J. Frazier and T. Galliard. The Royal Society of Chemistry, London, 199–215.
9. COLAS, A. and CHARGELEGUE, A. (1974). *Ann. Technol. Agric.*, **23**, 323–34.
10. MORRISON, W. R. and BARNES, P. J. (1983). In *Lipids in Cereal Technology*. Ed. P. J. Barnes. Academic Press, London, 149–63.
11. KENT, N. L., SIMPSON, A. G., JONES, C. R. and MORAN, T. (1944). *Milling*, **103**, 294–300.
12. KENT, N. L., THOMLINSON, J. and JONES, C. R. (1949). *Milling*, **113**, 46–54.
13. HINTON, J. J. C. (1944). *Biochem. J.*, **38**, 214–17.
14. STEVENS, D. J. (1959). *Cereal Chem.*, **36**, 452–61.
15. BARNES, P. J. and TAYLOR, P. W. (1980). *J. Sci. Food Agric.*, **31**, 997–1006.
16. MORRISON, W. R., COVENTRY, A. M. and BARNES, P. J. (1982). *J. Sci. Food Agric.*, **33**, 925–33.
17. MORRISON, W. R., WYLIE, L. J. and LAW, C. N. (1984). *J. Cereal Sci.*, **2**, 145–52.
18. MOSS, R., MURRAY, L. F. and STENVERT, N. L. (1984). *Bakers Digest*, **58**(3), 12–17.
19. ALLEN, J. C. and HAMILTON, R. J. (Eds) (1983). *Rancidity in Foods*. Elsevier Applied Science, London.
20. SHEARER, G. and WARWICK, M. J. (1983). In *Lipids in Cereal Technology*. Ed. P. J. Barnes. Academic Press, London, 253–67.
21. WARWICK, M. J. and SHEARER, G. (1982). *J. Sci. Food Agric.*, **33**, 918–24.
22. CUENDET, L. S., LARSON, E., NORRIS, C. G. and GEDDES, W. F. (1954). *Cereal Chem.*, **31**, 363–89.
23. BARNES, P. J. and LOWY, G. D. A. (1986). *J. Cereal Sci.*, **4**, 225–32.
24. GALLIARD, T. (1986). *J. Cereal Sci.*, **4**, 33–50.
25. GALLIARD, T. (1986). *J. Cereal Sci.*, **4**, 179–92.
26. CARLSON, T. L-G. (1981). Ph.D. thesis, University of Lund, Sweden.
27. MACRITCHIE, F. (1980). *Adv. Cereal Sci. Technol.*, **3**, 271–326.
28. SCHUSTER, G. and ADAMS, W. F. (1984). *Adv. Cereal Sci. Technol.*, **6**, 139–287.
29. CHUNG, O. K. and POMERANZ, Y. (1981). *Bakers Digest*, **55**(5), 38–50, 55, 96, 97.
30. FRAZIER, P. J. (1979). *Bakers Digest*, **53**(6), 8–10, 12, 13, 16, 18, 20, 29.
31. FRAZIER, P. J. (1983). In *Lipids in Cereal Technology*. Ed. P. J. Barnes. Academic Press, London, 189–212.

32. LARSSON, K. (1983). In *Lipids in Cereal Technology*. Ed. P. J. Barnes. Academic Press, London, 237–51.
33. LARSSON, K. (1986). In *The Chemistry and Physics of Baking*. Ed. J. M. V. Blanshard, P. J. Frazier and T. Galliard. The Royal Society of Chemistry, London, 62–74.
34. MACRITCHIE, F. (1983). In *Lipids in Cereal Technology*. Ed. P. J. Barnes. Academic Press, London, 165–88.
35. MACRITCHIE, F. (1984). *Adv. Food Res.*, **29**, 201–75.
36. POMERANZ, Y. (1980). *Bakers Digest*, **54**(1), 20–7; **54**(2), 12, 14, 16, 18, 20, 24, 25.
37. MACRITCHIE, F. and GRAS, P. W. (1973). *Cereal Chem.*, **50**, 292–302.
38. CHUNG, O. K., POMERANZ, Y. and FINNEY, K. F. (1982). *Cereal Chem.*, **59**, 14–20.
39. BUSHUK, W., BÉKÉS, F., MCMASTER, G. J. and ZAWISTOWSKA, U. (1984). In *Proc. 2nd Int. Workshop on Gluten Proteins*, Wageningen, 1984. Ed. A. Graveland and J. H. E. Moonen. TNO, Netherlands, 101–9.
40. BÉKÉS, F., ZAWISTOWSKA, U., ZILLMAN, R. R. and BUSHUK, W. (1986). *Cereal Chem.*, **63**, 327–31.
41. BELL, B. M., DANIELS, D. G. H., FEARN, T. and STEWART, B. A. (1987). *J. Cereal Sci.*, **5**, in press.
42. MACRITCHIE, F. (1983). In *Bread Research Institute, Wheat Research Unit Annual Report 1981/82, 1982/83*. CSIRO, Australia, 15.
43. CARLSON, T. L-G., LARSSON, K. and MIEZIS, Y. (1978). *Cereal Chem.* **55**, 168–79.
44. LARSSON, K. and PUANG-NGERN, S. (1979). In *Advances in the Biochemistry and Physiology of Plant Lipids*. Ed. L.-A. Appelqvist and C. Liljenberg. Elsevier/North Holland Biomedical Press, Amsterdam, 27–33.
45. POMERANZ, Y. (1971). *Baker's Digest*, **45**(1), 26–31, 58.
46. FRAZIER, P. J., DANIELS, N. W. R. and RUSSELL EGGITT, P. W. (1981). *J. Sci. Food Agric.*, **32**, 877–97.
47. BUSHUK, W. (1986). In *The Chemistry and Physics of Baking*. Ed. J. M. V. Blanshard, P. J. Frazier and T. Galliard. The Royal Society of Chemistry, London, 147–54.
48. BÉKÉS, F., ZAWISTOWSKA, U. and BUSHUK, W. (1983). *Cereal Chem.*, **60**, 371–8.
49. BÉKÉS, F., ZAWISTOWSKA, U. and BUSHUK, W. (1983). *Cereal Chem.*, **60**, 379–80.
50. ZAWISTOWSKA, U., BÉKÉS, F. and BUSHUK, W. (1984). *Cereal Chem.*, **61**, 527–31.
51. ZAWISTOWSKA, U., BÉKÉS, F. and BUSHUK, W. (1985). *Cereal Chem.*, **62**, 284–9.
52. FRAZIER, P. J., BRIMBLECOMBE, F. A. and DANIELS, N. W. R. (1974). *Proc. IV Int. Congress Food Sci. Technol.*, Vol. 1, 127–9.

7

Varietal Differences in Fatty Acid Compositions

R. J. HAMILTON
*Department of Chemistry and Biochemistry,
Liverpool Polytechnic, Liverpool, UK*

7.1 INTRODUCTION

Plant fatty acid compositions have been studied from a taxonomic point-of-view from the earliest days of their study during the 19th Century. Certainly Hilditch[1] in 1934 was suggesting that fats could be used in chemical taxonomy. Other workers[2] took the view propounded in 1956 by Erdtman[3] that the most valuable substances taxonomically are not those which are involved in primary metabolic processes but rather those which are relatively stable by-products. Despite these doubters, Hilditch,[4] in 1956, in his compendium of oils and fats, had begun to show how useful fatty acid compositions could be. Shorland[5] in 1963 summarised the position of fatty acids in taxonomy by stating 'that although the data on the types and distribution of fatty acids do not provide an unequivocal guide to the classification of plants, many correlations of taxonomic significance have become apparent in spite of the small number of species examined up to now'.

In the intervening two decades, many more plant species have been studied. The factors which can alter the fatty acid composition of a plant are slowly being recognised, viz. gene loci, temperature, humidity. Although it has not been possible to bring together the plant lipid compositions from twenty years, it was felt that it would be helpful to start with some of the work for the last five years. The compositions are listed for varieties in Tables 7.1–7.19, for varieties grown in different environmental conditions (Tables 7.20–7.32) and finally for some minor seed oils (Tables 7.33–7.51).

7.2 PLANT LIPID COMPOSITIONS

TABLE 7.1
EFFECT OF VARIETIES/SELECTIONS OF YELLOW-SEEDED RAI (*Brassica juncea*) ON OIL, PROTEIN, IODINE VALUE, ALLYL-ISOTHIOCYANATE AND REFRACTIVE INDEX OF THE OIL

Varieties/selection	Oil (%)	Protein (%) $N \times 6.25$	Iodine value	Allyl-isothiocyanate (Sinigrin) %	Refractive index at 26°C
YSK-2 (T-62)	31.6	38.9	105.4	0.79	1.4671
DYS-3 R 11	35.2	40.6	110.5	0.58	1.4674
YSK-3 (45-9)	44.1	34.1	114.3	0.59	1.4674
YSIK-7111	36.7	41.8	111.8	0.66	1.4674
YSK-1 (K-88)	31.2	42.9	110.5	0.64	1.4676
YSLK-74	25.5	41.8	113.0	0.68	1.4650
DYS-1	27.5	38.2	114.3	0.72	1.4671
YSP-408	22.4	44.6	114.3	0.43	1.4673
YSIK-742	38.4	41.8	109.2	0.78	1.4676
YSIK-4	27.8	45.9	111.8	0.53	1.4672
YSP-6	25.4	38.8	107.9	0.94	1.4672
YSP-9	26.4	41.8	105.4	0.67	1.4672
DYS-2	26.4	44.7	108.0	0.76	1.4671
PTYS-6	20.2	46.5	114.3	0.43	1.4673
Sl.1	42.1	40.8	111.8	0.76	1.4674
Sl.2	42.3	40.1	111.8	0.78	1.4677
Sl.3	39.1	42.2	105.4	0.62	1.4677

Kumar, R. and Abidi, A. B. (1981). *JAOCS*, **58**, 947.

Yellow mustard is an important oil in India. This paper is an attempt to report the quality parameters of mustard grown in Uttar Pradesh. Variety YSP-408 had the highest protein content.

TABLE 7.2
CHEMICAL CHARACTERISTICS OF MANGO FATS[2]

Variety	Fat content (%)	Iodine value	16:0	18:0	18:1	18:2	18:3	20:0
Brindaboni	9.3	51.2	9.2	39.1	41.1	7.6	0.7	2.3
Fazli	7.1	55.0	8.4	39.4	42.5	7.3	0.7	1.7
Kalabau	9.0	44.0	8.1	40.2	41.5	7.3	0.6	2.2
Kanchamitha	7.5	50.4	9.1	38.2	43.0	6.3	0.9	2.5
Kuipahari	8.0	50.3	10.0	38.3	41.3	7.1	0.9	2.5
Lakhanvoge	10.0	46.8	9.7	38.3	41.7	6.9	0.7	2.6
Lengra	9.7	46.0	8.5	39.3	42.4	6.8	0.7	2.2
Mohanvoge	7.8	55.0	9.0	38.6	41.4	7.6	0.9	2.4
Misrakanta	8.9	50.6	7.9	39.2	43.8	6.0	1.0	2.0
Ranipasand	9.4	51.2	7.9	39.9	42.0	7.4	0.6	2.2

Ali, M. A., Gafur, M. A., Rahman, M. S. and Ahmed, G. M. (1985). *JAOCS*, **62**, 520.

Bangladesh is a major mango growing area. The fat content of mango kernels grown under soil and climatic conditions in Bangladesh varied from 7.1% to 10% depending on the variety. Sterols and phospholipids were also analysed.

TABLE 7.3
OIL CONTENT, IODINE VALUE AND FATTY ACID COMPOSITION OF COTTONSEED VARIETIES

Cottonseed varieties	Oil content (%)	Iodine value	14:0	14:1	16:0	16:1	18:0	18:1	18:2	20:0
B.S.1.	24·0	107·2	1·4	—	28·8	2·4	3·5	22·5	41·2	0·2
B-557	24·0	103·6	0·9		44·9	—	0·3	16·2	37·7	—
149-F	23·0	106·9	0·9		30·6	1·8	2·3	23·4	40·7	—
AC-134	21·8	104·8	2·0		43·1	2·7	0·5	17·4	34·3	—
K-68-9	21·7	105·3	1·1		33·9	0·9	3·1	27·5	33·3	0·2
Qalandri	16·8	105·0	0·7		32·2	0·3	2·2	27·8	36·8	—
AC-Sekrand	16·0	102·9	1·3		36·5	—	8·5	22·1	31·3	0·3
D-9-Sekrand	11·6	105·9	0·7		33·1	0·5	1·4	32·0	32·0	0·3

Raie, M. Y., Ahmad, M., Ahmad, I., Khan, S. A. and Athar Jafri, S. A. (1983). *Fette Seifen Anstrichmittel*, **85**, 279.
Eight varieties of cottonseed oils grown in Pakistan are studied.

TABLE 7.4
OIL CONTENT, IODINE VALUE AND FATTY ACID COMPOSITION OF HYBRID COTTONSEED VARIETIES

Hybrid cotton seed	Oil content (%)	Iodine Value	14:0	16:0	18:0	18:1	18:2	20:0	22:0	CPA[a]
DCH-32	23·8	92·9	1·1	34·1	4·5	10·3	46·7	1·1	1·0	1·2
DS-56	22·5	95·8	1·9	32·1	4·4	9·7	48·6	1·1	1·0	1·2
DS-59	24·5	96·4	3·0	25·2	3·8	22·1	42·4	1·1	1·2	1·2
SRG-26	22·7	101·3	1·7	24·3	3·6	20·4	46·2	1·4	1·1	1·3
DS-4	23·1	102·3	1·6	25·5	5·2	14·4	49·9	1·2	1·2	1·0
DP-197	24·1	102·5	1·9	26·7	2·6	17·2	48·9	0·8	1·0	0·9
DP-225	24·5	105·1	1·7	24·0	2·2	20·7	48·8	0·8	1·0	0·8
CPD-8-1	23·9	106·7	1·7	21·5	4·0	21·8	48·6	0·8	0·9	0·7
DCH-295	24·0	111·4	1·0	21·2	3·9	18·8	52·5	0·8	0·9	0·9
DCH-337	24·3	114·0	0·9	18·3	3·2	23·4	51·5	0·9	1·0	1·2

[a]Cyclopropenoid acid.
Badami, R. C., Alagawadi, K. R. and Shivamarthy, S. C. (1982). *Fette Seifen Anstrichmittel*, **84**, 278.

TABLE 7.5
OIL CONTENT, IODINE NUMBER AND FATTY ACID COMPOSITION OF SOYABEAN VARIETIES

Soyabean varieties.	Oil content (%)	Iodine number	14:0	16:0	18:0	18:1	18:2	18:3	20:0	22:0	24:0
Callend 'K'	17·5	102·4	0·2	13·9	4·7	24·5	46·9	8·6	0·4	0·3	0·5
Bragg 'K'	15·9	99·7	0·3	13·8	5·4	31·8	41·2	6·3	0·6	0·2	0·4
Loppa 'S'	17·1	106·8	0·4	13·7	3·2	21·2	50·6	9·2	0·6	0·6	0·5
Callend 'S'	19·1	124·3	0·2	11·7	3·3	31·6	47·6	5·1	0·2	0·2	0·3
Ford 'S'	20·1	126·2	0·3	11·9	1·2	26·1	53·4	5·8	0·3	0·6	0·4
Beruit No. 38 'S'	16·9	117·5	0·3	12·6	2·1	26·0	48·7	8·8	0·2	0·4	0·3

K stands for Kharif (Autumn) and S for Spring crops.
Raie, M. Y. and Iqbal, M. L. (1983). *Fette Seifen Anstrichmittel*, **85**, 194.
Cottonseed and rape/mustard seed are the major oil seeds of Pakistan. Attempts are being made to enhance the production of these crops and to introduce other oil seed crops to Pakistan. The oil seeds, which have been analysed, were collected in the USA and found to be very successful as Kharif (Autumn) crops in the rainfed area of Pakistan. These seeds have a higher yield of oil.

TABLE 7.6
FATTY ACID COMPOSITION OF MAIZE GROWN IN RUSSIA

(a) Hybrids of yellow corn mother plant

Hybrid	Genotype	Year	16:0	18:0	18:1	18:2	18:3
B × A	YYy	1978	11·0	1·4	35·3	51·3	1·0
B × A	YYy	1980	10·4	1·3	30·3	56·0	2·0
B × C	YYy	1978	11·1	1·2	30·2	54·4	3·1
B × C	YYy	1980	11·2	1·3	30·2	55·7	1·6
D × A	YYy	1978	11·5	1·1	27·8	57·6	2·0
D × A	YYy	1980	11·5	1·5	31·8	53·3	1·8
F × A	YYy	1978	11·3	1·4	29·3	56·7	1·4
F × A	YYy	1980	9·5	1·7	32·0	52·2	4·6
F × D	YYY	1978	11·4	2·4	27·0	58·1	1·1
F × D	YYY	1980	10·7	1·2	30·3	55·2	2·5
D × B	YYY	1978	11·8	1·5	30·4	54·4	1·9
D × B	YYY	1980	11·9	1·3	29·1	55·8	1·9
B × E	YYY	1978	11·5	1·1	27·8	57·6	2·0
B × E	YYY	1980	12·4	1·3	25·3	58·6	2·4
B × D	YYY	1978	11·4	1·5	29·9	55·8	1·5
B × D	YYY	1980	10·9	1·2	27·6	51·7	2·6
E × A	YYy	1978	14·1	1·3	26·7	56·9	1·0
E × A	YYy	1980	14·0	1·1	26·9	56·0	2·0
E × B	YYY	1978	12·7	1·7	23·0	60·0	2·6
E × B	YYY	1980	13·3	1·1	26·4	57·2	2·0
E × C	YYy	1978	12·0	1·1	23·7	61·5	1·6
E × C	YYy	1980	13·5	1·2	22·5	60·8	2·0
D × E	YYY	1978	11·5	1·7	27·8	58·3	1·9
D × E	YYY	1980	13·5	1·3	27·3	56·1	1·8

Karaiwanov, G., Marquard, R. and Petrovskij, E.W. (1982). *Fette Seifen Anstrichmittel*, **84**, 251.

(b) Hybrids from white corn

Hybrid	Genotype	Year	16:0	18:0	18:1	18:2	18:3
A × C	yyy	1978	11·2	1·6	28·0	57·4	1·7
A × C	yyy	1980	11·9	1·2	25·7	59·5	1·7
C × A	yyy	1978	12·2	1·2	25·4	59·0	2·2
C × A	yyy	1980	10·4	1·2	23·3	63·1	2·0
A × B	yyY	1978	10·7	1·7	34·7	51·8	1·2
A × B	yyY	1980	10·9	1·4	32·3	53·2	2·2
A × D	yyY	1978	12·4	1·8	32·4	52·2	1·2
A × D	yyY	1980	12·2	1·4	30·3	54·6	1·5
C × B	yyY	1978	12·4	1·6	28·6	56·1	1·4
C × B	yyY	1980	10·3	1·1	26·4	58·7	3·5
A × E	yyY	1978	14·5	1·3	27·6	54·8	1·9
A × E	yyY	1980	14·0	1·1	26·9	56·0	2·0

Karaiwanov, G., Marquard, R. and Petrovskij, E.W. (1982). *Fette Seifen Anstrichmittel*, **84**, 251.

TABLE 7.6—contd.

(c) Maize lines

Line	Genotype	Year	16:0	18:0	18:1	18:2	18:3
MKK 18 white	yyy	1978	12·0	1·9	34·3	47·7	2·1
MKK 18 white	yyy	1980	10·8	1·3	28·6	57·4	1·9
MKK 18 yellow	YYY	1978	9·8	1·6	32·2	54·2	2·3
MKK 18 yellow	YYY	1980	10·6	1·4	26·0	60·3	1·7
MKK 42 white	yyy	1978	12·3	1·3	21·0	64·2	1·2
MKK 42 white	yyy	1980	12·5	1·1	19·4	64·9	2·1
MKK 50 yellow/red	YYY	1978	12·0	1·7	30·1	54·7	1·6
MKK 31 red/yellow	YYY	1978	16·4	1·1	22·0	58·5	2·0
MKK 31 red/yellow	YYY	1980	12·9	1·2	28·8	54·8	2·2
MKK 776 yellow	YYY	1978	11·0	1·4	30·8	55·5	1·4
MKK 776 yellow	YYY	1980	11·5	1·2	29·8	54·9	2·6

Kardiwanov, G., Marquard, R. and Petrovskij, E. W. (1982). *Fette Seifen Anstrichmittel*, **84**, 251.

These Russian maize lines and maize hybrids show that fatty acid composition is mainly directed by male genes.

TABLE 7.7
MOISTURE, PROTEIN AND LIPID LEVELS WITH FATTY ACID COMPOSITIONS OF VARIETIES OF OATS *Avena sativa*

Oat Strain	Moisture %	Protein %	Lipid %	14:0	16:0	18:0	18:1	18:2	18:3	Other
Exeter	8·4	13·2	4·2	1·0	25·8	2·9	25·8	40·6	3·7	0·2
Garry	7·7	20·4	5·4	0·6	22·4	2·6	30·0	41·0	3·3	0·1
Hinoat	7·4	24·4	5·5	3·1	19·6	2·5	32·8	36·9	2·8	2·2
OA-290-5	7·8	18·8	5·5	3·6	23·1	3·0	39·8	38·0	2·5	d
Gemini	7·4	16·4	5·9	1·1	21·3	2·0	32·0	40·3	3·2	0·1
Random	8·0	17·9	6·3	0·5	21·0	2·2	36·1	37·4	2·4	0·3
Terra	7·7	16·2	6·5	4·9	19·7	2·4	35·6	32·3	2·0	3·1
Elgin	7·8	19·3	6·8	1·0	23·0	3·9	30·8	37·6	3·6	+
Dal	7·8	19·1	7·9	0·5	23·6	2·8	33·7	37·7	1·7	+
Cl 3387	7·7	14·1	11·0	2·7	16·5	1·6	40·2	35·5	2·2	1·3
Lodi	7·8	19·9	11·3	1·4	14·9	2·4	41·0	36·0	2·1	2·2
Cl-4492	8·0	14·2	11·8	4·9	15·6	2·7	41·3	31·3	1·7	2·4

6ft × $\frac{1}{8}$" 10% SP22P2P on 100–20 mesh.
Sahasrabudhe, M. R. (1979). *JAOCS* **56**, 80.

TABLE 7.8
STEROL AND FATTY ACID COMPOSITION FOR ALMOND, PEACH AND APRICOT VARIETIES

	14:0	16:0	16:1	17:0	17:1	18:0	18:1	18:2	18:3	20:0	20:1
Sweet almond	t	6·3	0·4	0·03	0·07	1·5	74·1	17·3	—	0·08	0·08
Peach variety Redhaven	0·01	5·2	0·3	0·04	0·08	3·3	78·5	12·2	t	0·11	0·08
Peach mixed variety	t	4·7	0·8	0·04	0·13	2·34	67·3	24·3	0·05	0·10	0·06
Apricot variety Bulida	t	4·8	0·7	0·08	0·16	0·96	66·8	25·9	0·11	0·06	0·14
Apricot mixed variety	t	7·0	0·3	0·07	0·09	2·7	58·3	31·0	0·07	0·13	0·08

Cap. col. 20 m × 0·3 mm. Coated with EGA. Temperature 170°C.
Salvo, F., Dugo, G., Stagno D'Alcontres, I., Cotroneo, A. and Dugo, G. (1980). *La Riv. Ital. Sost. Grasse*, **LVII**, 24.
This study was conducted to distinguish between sweet almond oil, peach and apricot seed oils on the basis of both fatty acid composition and the sterol composition.

	Cholesterol	Campesterol	Stigmasterol	B-Sitosterol	Δ^3-Avenasterol	Δ^7-Stigmasterol	?	Δ^7-Avenasterol	?
Sweet almond	0·19	2·7	0·8	82·9	11·4	t	0·7	0·4	0·7
Peach variety Redhaven	0·09	3·1	0·6	88·1	5·7	0·25	0·5	0·3	1·2
Peach mixed variety	t	3·8	0·3	89·4	5·5	0·05	0·1	0·1	0·4
Apricot variety Bulida	t	3·4	0·2	91·6	3·7	t	0·2	0·1	0·4
Apricot mixed variety	t	4·2	0·5	90·0	3·1	0·20	0·2	0·2	1·3

Sterols as TMSi on OV17 30 m × 0·3 mm at 240°C.

TABLE 7.9
RAPESEED MUTANTS

Mutant group	Genotype	18:2	18:3	Ratio 18:3 18:2
0	TOWER	22·8	9·7	0·42
	ORO	21·5	9·8	0·45
	EGRA	20·3	10·1	0·50
I	M57	22·4	5·6	0·25
II	M3	32·5	7·4	0·22
	M6	23·4	3·5	0·15
	M8	33·8	8·3	0·24
	M11	37·8	8·4	0·22
III	M40	30·5	5·4	0·17
	M41	26·1	3·8	0·15
	M42	24·6	3·5	0·15
	M43	27·7	3·3	0·12
	M44	29·1	4·5	0·15
	M45	28·3	4·0	0·14
	M46	32·4	4·5	0·14
	M47	30·1	3·2	0·11

Groups I and II are one step mutation of '010' seeds.
Mutant group III are derived from further mutation of mutant M57.
Ratledge, C. (1984). *Fette Seifen Anstrichmittel*, **86**, 379.

TABLE 7.10
FATTY ACID COMPOSITION OF PEANUT VARIETIES

		16:0	18:0	18:1	18:2	20:0	20:1	22:0	24:0
Peanut	Var 37	12·5	4·5	34·9	39·6	2·2	1·0	4·0	1·4
	Var 50	11·4	2·3	43·3	37·2	1·1	1·2	2·5	1·0
	Var 48·1[a]	10·2	1·5	52·3	27·7	1·0	2·0	3·4	1·9
	Var 48·2[a]	12·2	1·3	44·2	33·8	1·0	1·9	3·7	2·0
	Var 54	7·2	3·5	71·2	12·3	1·7	1·0	2·2	0·8
	Var 80	12·2	3·7	39·9	37·1	1·8	0·9	3·3	1·2
	Var 58(1)[a]	6·5	3·4	73·3	12·0	1·5	0·8	1·8	0·8
	Var 58(2)[a]	8·3	3·3	62·2	20·6	1·5	1·0	2·3	0·9
	Var 60	7·8	3·8	68·0	14·9	1·6	1·0	2·2	0·9
Arachis vittosullarpa		9·9	1·7	16·2	51·6	1·5	2·2	12·5	4·4

[a]Seed from different crop years.
1·86 m × 4 mm 15% EGS on Chromosorb W(AW)DMCS.
Hopes, J. C. and Worthington, R. E. (1979). *JAOCS*, **56**, 953.
Triacylglycerols were isolated from peanut cultivars chosen to include known extremes in oleic and linoleic acid content. Molar concentrations of oleic and linoleic acid in *sn* 2-monoacylglycerols were highly correlated with molar concentrations in triacylglycerols.

TABLE 7.11
COMPOSITIONS OF FATTY ACIDS AND TRIACYLGLYCEROLS FROM PEANUTS GROWN IN DIFFERENT GEOGRAPHICAL SITUATIONS IN AFRICA

	Peanut varieties		16:0	18:0	18:1	18:2	20:0	20:1	22:0	24:0
Zaire	A65		12.7	2.6	38.8	29.4	1.8	1.3	2.4	1.0
	A20		11.9	3.8	39.8	37.1	2.4	1.5	2.6	0.9
	A1052		10.8	4.4	42.7	35.6	2.4	1.0	2.7	1.0
	A1055		10.7	4.7	40.6	36.0	2.7	1.7	2.9	0.7
	P43		11.4	4.8	40.2	35.8	2.7	1.2	3.0	0.9
Senegal	Bambey 55–437		11.3	3.0	49.5	31.0	1.2	0.8	2.6	0.6
	Bambey 28–206		9.1	2.1	66.2	17.2	1.2	1.4	1.9	0.9
	Bambey 73–33	1	10.6	2.2	58.9	22.9	1.4	1.2	1.9	0.9
	A65	2	22.0	2.0	39.6	33.2	1.3	0.6	0.8	0.5
		3	0.9	0.2	30.3	68.3	—	0.3	—	—
			16.7	5.7	47.5	16.2	4.2	2.4	6.1	1.2
	A20	1	20.4	6.3	37.2	31.5	1.6	0.8	1.7	0.5
		2	1.3	0.4	37.0	61.3	—	—	—	—
		3	15.0	5.8	43.6	18.3	5.8	3.1	6.2	2.2
	A1052	1	17.6	7.0	39.6	30.8	1.7	0.7	1.9	0.7
		2	0.7	—	36.2	63.1	—	—	—	—
		3	14.4	6.8	48.3	12.5	6.1	2.7	6.9	2.3
	A1055	1	17.6	7.3	38.2	31.3	2.1	1.0	1.9	0.6
		2	0.7	0.1	36.4	62.8	—	—	—	—
		3	13.5	7.4	45.6	13.6	6.5	3.6	7.3	2.5

(continued)

TABLE 7.11—contd.

Peanut varieties		16:0	18:0	18:1	18:2	20:0	20:1	22:0	24:0
P43	1	17·9	7·7	40·1	28·9	1·6	0·6	2·5	0·7
	2	1·2	0·2	36·5	62·1	—	—	—	—
	3	14·5	7·0	44·2	17·1	6·2	2·5	6·9	1·6
Bambey 55–437	1	15·4	3·4	52·4	25·6	0·6	0·4	2·0	0·2
	2	1·1	—	44·2	54·7	—	—	—	—
	3	15·6	4·6	53·0	15·8	3·0	1·1	5·8	1·1
Bambey 28–206	1	15·6	4·0	64·4	10·9	0·7	1·1	1·8	1·5
	2	1·3	0·2	69·2	29·3	—	—	—	—
	3	11·1	2·1	67·8	8·2	2·9	3·1	3·6	1·2
Bambey 73–33	1	15·6	3·4	59·9	18·1	0·5	0·9	1·2	0·4
	2	2·0	0·5	60·0	37·5	—	—	—	—
	3	14·0	3·4	58·7	12·3	3·4	2·6	3·8	1·8

1·82 m × 3 mm 10% SP 2330 on 100–200 mesh. 190–250°C, Chromosorb AW, 8° per min. Van Pee, W., Van Hee, J., Boni, L. and Hendrikx, A. (1979). *JAOCS*, **56**, 901.
Peanut varieties from two different locations were analysed and the total fatty acids are shown in first half of table. The fatty acids at positions 1, 2 and 3 have also been examined.

TABLE 7.12
TURKISH GROWN PEANUT VARIETIES

Variety		16:0	18:0	18:1	18:2	20:0	22:0	24:0
Tamnut 74		10·7	2·8	39·9	41·4	1·2	1·2	2·8
Florunner		10·7	2·4	44·2	38·6	0·6	1·2	2·2
Senegal 57		10·4	2·4	38·9	44·1	0·8	1·3	2·1
Okla. P		11·0	2·2	43·1	39·3	0·8	0·9	2·7
Starr (SP)		11·7	2·0	44·2	37·6	1·3	1·2	2·0
Erzin Selected	4/4	10·6	1·9	46·3	37·2	0·9	1·4	1·7
	6/6	11·6	2·5	45·9	36·4	0·9	1·2	1·6
	7/4	11·4	2·1	45·0	37·5	0·8	1·3	1·9
	8/9	12·8	2·7	47·6	33·3	1·0	1·3	1·3
	11/3	11·1	2·7	45·3	37·1	0·9	1·2	2·0
	12/4	10·3	2·7	42·4	40·4	0·9	1·3	2·1
	14/2	11·2	1·1	45·9	37·4	0·7	1·0	2·6
	19/1	11·3	1·7	46·9	38·8	0·7	0·7	<0·2
	22/5	10·9	1·4	47·2	37·0	0·8	1·1	1·6

Emiroglu, S. H. and Marquard, R. (1984). *Fette Seifen Anstrichmittel*, **86**, 103.
Peanut breeding programme at Turkish University. Large grain selected genotypes were investigated and compared with international varieties from the same cultivation place. There was a wide variation in productivity 15 dt/ha. A mean of 2 years by variety 'Tamnut 74' and 'Starr'. The thousand grain weights are on average 50% higher than the average of the international variety. These genotypes are lower in fat content than the international varieties.

TABLE 7.13
FATTY ACID COMPOSITION OF 20 CASSAVA CULTIVARS (*Manihot esculenta* Crantz)

Variety	% Total lipids	16:0	17:0	18:1	18:2	18:3	% Dry matter crude protein
Sem Nome	0·83	39·8	8·1	49·2	2·9	—	2·61
Manteigas	1·04	32·8	7·0	50·1	3·8	6·2	1·39
SFG 2317	2·01	56·6	—	36·6	—	6·8	2·29
SFG 469	2·03	47·3	—	52·7	—	—	1·98
Saracura	0·70	58·0	3·2	24·6	12·6	1·5	2·47
Mangue mirim	1·30	32·4	6·5	48·2	11·3	1·6	2·34
Mawana[a]	1·14	33·2	—	46·0	15·3	4·7	1·44
JL-8[b]	1·28	32·4	6·6	40·1	14·3	5·6	2·47
Roxinha[c]	0·90	37·8	7·3	41·8	9·2	2·7	3·64
Caravela	1·52	44·9	2·9	46·1	4·0	2·2	1·84
Prato	1·72	28·0	3·5	46·3	15·2	6·4	2·34
Santinha	0·80	31·8	1·8	41·9	19·0	5·6	3·92
Variedade I	1·23	31·7	1·7	47·6	13·5	5·5	2·47
Veada	1·42	36·6	—	50·5	12·9	—	3·10
Amargosa	0·92	38·6	—	49·0	12·4	—	1·70
Ligeirinha	1·32	38·6	1·4	51·0	7·1	1·9	4·10
Gostosa[d]	1·84	41·2	2·4	43·2	8·9	4·3	4·10
Vermelhinha[e]	0·92	33·1	2·4	43·1	11·6	7·9	4·70

In addition, variety (a) has 0·80% $C_{14:0}$; (b) has 0·96% $C_{14:0}$; (c) has 1·02% $C_{14:0}$; (d) has 0·1% $C_{14:0}$; (e) has 2·0% $C_{14:0}$.

Teles, F. F. F., Oliveira, J. S., Silveria, A. J., Batista, C. M. and Stull, J. W. (1985). *JAOCS*, **62**, 70.

Twenty cultivars of Cassava from Brazil have been analysed for fatty acids, carbohydrates and protein.

TABLE 7.14
OIL CONTENT AND FATTY ACID COMPOSITION OF PEANUT VARIETIES

Peanut variety	Oil content (%)	16:0	18:0	18:1	18:2	20:0	20:1	22:0	24:0
Giza 1	48·4	12·7	2·6	38·8	39·4	1·8	1·3	2·4	1·0
Giza 2	48·2	11·9	3·8	39·8	37·1	2·4	1·5	2·6	1·0
Giza 3	48·6	11·4	4·8	40·2	35·8	2·7	1·2	3·0	1·0

Abdel-Hamid, Y. and Abdel-Rahman, J. (1982). *JAOCS*, **59**, 287.

TABLE 7.15

Safflower species	Oil content %	Iodine value	14:0	16:0	18:0	18:1	18:2
U.S. -10	32·0	128·6	3·1	10·2	5·5	14·4	66·8
V.F. Stp (53-1)	30·5	130·2	2·8	12·0	3·6	15·7	65·9
S-208	29·7	142·4	0·9	9·4	2·3	14·0	73·4

Raie, M. Y., Muhammad, D. and Khan, S. A. (1985). *Fette Seifen Anstrichmittel*, **87**, 282.

TABLE 7.16
SOYA VARIETIES GROWN IN SICILY

Soya varieties	14:0	16:0	18:0	18:1	18:2	18:3
Pella	0·1	13·3	4·8	22·1	52·8	6·8
Columbus		11·4	5·3	25·3	51·5	6·5
Amsoy	0·1	14·0	3·9	26·5	51·4	4·1
Hodgson	0·1	12·4	3·8	21·9	53·4	8·4

Panno, M. and Palazzolo, E. (1984). *Fette Seifen Anstrichmittel*, **61**, 497.

TABLE 7.17
FATTY ACID COMPOSITION OF PEANUT VARIETIES

Peanut varieties	Position	16:0	18:0	18:1	18:2	20:0	20:1	22:0	24:0
Flori giant	TG	10.8	2.9	53.1	27.3	1.8	1.0	1.9	1.1
	1	20.1	4.9	50.7	22.6	0.5	0.7	0.4	0.3
	2	2.2	0.7	51.5	45.3	0.1	0.3	0.1	—
	3	10.3	3.2	57.2	14.0	4.8	2.0	5.3	3.0
Early bunch	TG	13.3	2.6	42.0	36.3	1.5	0.9	2.4	1.1
	1	24.7	4.4	38.3	31.2	0.3	0.4	0.5	0.2
	2	3.5	1.5	37.2	57.4	0.1	0.2	—	—
	3	11.8	1.9	50.6	20.3	4.0	1.9	6.5	2.9
Florunner	TG	11.4	2.1	50.9	29.1	1.6	1.1	2.4	1.3
	1	20.7	3.5	49.5	24.4	0.3	0.7	0.5	0.5
	2	2.1	0.6	47.8	48.8	0.1	0.4	0.1	0.1
	3	11.4	2.3	55.5	14.1	4.4	2.3	6.5	3.4
Tifrun	TG	12.6	2.4	42.4	36.8	1.5	0.9	2.6	0.8
	1	22.7	4.2	38.8	32.7	0.3	0.4	0.6	0.2
	2	3.2	1.2	36.1	58.8	0.2	0.2	0.1	—
	3	11.9	1.7	52.2	18.9	4.0	2.0	7.2	2.2
Starr	TG	14.2	3.3	43.3	33.0	1.8	1.1	2.7	0.7
	1	24.2	4.9	40.4	28.4	0.4	0.6	0.7	0.3
	2	2.4	0.8	39.5	56.9	0.1	0.2	0.1	—
	3	16.0	4.2	50.0	13.8	4.8	2.4	7.3	1.8
Spancross	TG	13.5	2.9	44.1	33.4	1.9	1.1	2.2	0.9
	1	24.2	5.2	40.3	28.4	0.3	0.5	0.6	0.4
	2	2.6	0.9	40.2	56.2	0.1	—	—	—
	3	13.5	2.8	51.8	15.6	5.2	2.5	6.5	2.3

GLC on 10% EGSS-X on 100–120 Gas Chrom P.
Saunders, T. H. (1979). *Lipids*, **14**, 630.

TABLE 7.18
EFFECT OF CROSSES ON FATTY ACID COMPOSITION OF PALM

Elaeis Palm varieties	12:0	14:0	16:0	18:0	18:1	18:2	20:0	18:3
G, *Elaeis guineensis* unknown variety	0·1	1·3	47·0	3·3	39·7	8·4	0·1	0·1
M, *Elaeis oleifera*		0·2	21·0	0·2	58·3	20·0		0·3
F_1, first filial generation (*E. guineensis* × *E. oleifera*)		0·8	37·1	1·5	52·2	8·3	0·1	0·2
$F_1 \times G$		0·6	41·4	0·9	48·2	8·9		0·1
$F_1 \times M$		0·5	36·2	0·4	52·8	9·3		0·7
F_2, second filial generation ($F_1 \times F_1$)		0·4	39·1	0·7	53·3	6·5	0·1	0·1

2 m × 4 mm 10% DEGS on Chromosorb B, A/W DMCS.
Ong, S. H., Chuah, G. C. and Sow, H. P. (1981). *JAOCS*, **58**, 1032.

TABLE 7.19
FATTY ACID COMPOSITION OF RAPESEED LINES

	16:0	18:1	18:2	18:3
1 Herkules	5	58	23	12
2 F_1	10	30	47	9
3	11	35	44	11
4	5	58	33	3
5	5	76	11	6

Robbelen, G. (1983). *Fette Seifen Anstrichmittel*, **85**, 395.

7.3 VARIETIES GROWN FOR DIFFERENT ENVIRONMENTAL CONDITIONS

TABLE 7.20
FATTY ACID COMPOSITION OF MOWRAH FAT FROM SEVERAL DISTRICTS IN WEST BENGAL.

West Bengal Districts	14:0	16:0	18:0	18:1	18:2	18:3	20:0
Sonamukhi Bankura	—	23.0	23.0	38.8	14.8	—	0.5
Sunuk Pahari	0.1	27.8	20.0	36.3	15.4	—	0.4
Chittaranjan Burdwan	0.1	15.4	20.7	40.8	21.9	0.4	0.7
Bihar Districts							
Ranchi	0.1	21.0	22.6	39.7	16.0	—	0.6
Hazaribagh	0.3	18.4	20.2	45.6	15.3	—	0.3
Dalton Junj.	0.2	20.0	22.8	39.9	16.7	—	0.5
Orissa Districts							
Nowrangpur Papadahandi	—	23.4	15.9	41.9	17.8	0.4	0.7
Sambalpur	0.1	21.2	26.3	34.7	16.2	0.5	1.0
Baripada	0.1	22.4	25.3	37.0	13.9	0.3	0.9
Keonjhar	0.2	31.7	17.3	35.7	14.7	—	0.4
Jeypore	0.2	25.1	25.6	32.5	16.5	—	0.3
Rayagada	—	20.7	20.7	40.9	16.4	0.4	0.8
Madhya Pradesh							
Sahdol	—	22.0	24.1	36.0	16.9	0.3	0.8
Raigarh	0.1	22.5	24.7	34.7	17.0	0.2	0.8

15% DEGS on Chromosorb W HP. Mowrah fat from kernels of *Madhuca latifolia*.
Sen Gupta, S., Chakrabarty, M. M. and Bhattacharyya, D. K. (1982). *Fette Seifen Anstrichmittel*, **84**, 226.
Mowrah fat has become very important in India for the preparation of plastic fats like cocoa butter substitute, bakery fat and vanaspati. The unique fatty acid composition has led to this recent importance. This paper notes, for the first time, the variation in composition with the region of growth. Palmitic acid content has been shown to range from 15% to as high as 32%.

TABLE 7.21
FATTY ACID COMPOSITION OF A SELECTION OF SEEDS GROWN IN TURKEY

Turkish grown oil seeds	14:0	16:0	16:1	18:0	18:1	18:2	18:3	20:0	22:0	24:0
Groundnut	t	12·1	t	3·0	44·4	32·4	—	2·5	2·3	3·4
Sesame seed		10·5	0·1	6·3	36·3	45·7	—	1·2		
Poppyseed		9·8	t	1·3	14·9	73·9				
Hempseed		9·4		3·2	15·0	49·3	23·1			
Linseed		7·3		5·4	9·3	15·3	52·6			
Tobacco seed		8·8			13·7	75·3	1·6			
Syrian Scabious	18·4	8·8		1·9	33·3	36·3				
Tomato seed	0·1	16·1		5·3	20·8	55·8		1·9		
Grape seed	0·2	8·7		3·7	19·4	68·0				
Tea seed	0·1	16·2		1·3	61·4	19·9		1·1		
Laurelberry[a]	1·8	17·5	1·6	1·9	40·6	23·2	1·9			
Fig seed		7·2		2·6	14·9	30·6	44·7			
Peach kernel		8·0		0·3	55·1	36·5				

[a]Contains 11·1% of C_{12} and shorter chain acids.

Yazicioglu, T. and Karaali, A., (1983). *Fette Seifen Anstrichmittel*, **85**, 23.

The varying climatic conditions found in Turkey result in the cultivation of a wide variety of oil bearing crops, trees and nuts. The main oil seeds used in Turkey are olive, sunflower seeds and cotton seeds. They yield 97% of the edible oil consumed whilst the remaining 3% is made up of poppy, linseed, hempseed, rapeseed, soyabeans, safflower seeds and groundnuts. The fatty acids of some of these oils are reported. There is a third group, the non-traditional oil raw materials, tobacco seed, grape seed, scabious, fig seed, tomato seed, laurel seed, tea seeds and peach kernels whose compositions are also shown.

The values of the fatty acid composition for Turkish grown seeds are compared with literature values.

TABLE 7.22
FATTY ACID COMPOSITION OF RUBBER SEED OIL GROWN IN SEVERAL DISTRICTS IN INDIA

Rubber seed oil *Hevea brasiliensis*	Sap. value	Iodine value	8:0	9:0	10:1	11:0	14:0	14:1	16:0	18:0	18:1	18:2	18:3
Kadamkal, Sampaje Div. 1968	186	141		0·3	0·5	0·3	0·2		11·4	8·2	21·4	37·6	20·1
Kavu, Paittur Div. 1970	185	134	0·1		0·2	0·1	0·1		9·4	9·3	27·1	38·9	15·0
Medinadka, Sullia Div. 1969	186	141			t		0·1		10·0	5·8	27·5	41·6	15·0
Kerala (Commercial House)	186	135			0·6			0·4	11·0	9·4	23·0	41·0	14·6

Jayappa, V., Shanbhag, P. K., Amminally, S. and Patel, K. N. (1983). *Fette Seifen Anstrichmittel*, **85**, 472. Rubber seeds constitute 25–30% oil which is used as a soap stock. These oils come from six rubber plantations in Dakshina Kannada District of Karnataka State. This is the first report of caprylic and myristic acids as well as nonanoic and undecanoic acids.

TABLE 7.23

PROPORTIONS OF OLEIC AND LINOLEIC ACID AND STEROL COMPOSITIONS FOR SUNFLOWER SEED OIL GROWN IN DIFFERENT LOCATIONS IN ITALY

	18:1	18:2	Campesterol	Stigma-sterol	β-sit-oster-ol	$\Delta 7$ stigma-sterol
Bologna I	37·7	51·8	9·7	13·1	70·8	5·2
Bologna II	37·6	52·8	10·4	11·7	72·5	4·2
Bologna III	29·4	61·0	11·6	10·1	72·7	4·3
Pisa I	26·8	62·6	9·5	10·3	72·1	4·8
Pisa II	28·1	62·4	10·5	11·1	74·0	2·4
Pisa III	25·4	62·2	8·6	9·8	68·0	9·8
Osimo	33·7	56·7	10·0	9·5	73·4	5·4
Perugia I	24·7	64·0	8·6	11·8	73·3	3·9
Perugia II	26·2	61·8	11·9	10·4	67·3	6·6
Sassari	28·0	61·2	9·9	9·8	73·1	4·9
Battipaglia	30·8	58·6	9·4	10·1	74·4	3·1
Catania I	37·5	52·5	10·3	9·6	74·4	3·4
Catania II	34·4	55·2	10·5	12·2	72·6	3·7

Conte, L. S., Antonelli, A., Guglielmi, A. and Capella, P. (1984). *Riv. Ital. Sost. Grasse*, **61**, 481.

TABLE 7.24

FATTY ACID COMPOSITION OF SOYABEANS GROWN IN SICILY

Soyabean varieties	14:0	16:0	18:0	18:1	18:2	18:3
Pella	0·1	13·4	4·8	22·1	52·8	6·8
Columbus	—	11·4	5·3	25·3	51·5	6·5
Amsoy	0·1	14·0	3·9	26·5	51·4	4·1
Hodgson	0·1	12·4	3·8	21·9	53·4	8·4

Pamo, M. and Palazzolo, E. (1984). *Riv. Ital. Sost. Grasse*, **61**, 497.

TABLE 7.25
YEARLY VARIATION IN EARLY BUNCH TRIACYLGLYCEROL OF PEANUTS GROWN IN TWO LOCATIONS

Year	Location		Fatty acid distribution (mol %)							
			16:0	18:0	18:1	18:2	20:1	20:1	22:0	24:0
1975	G	TG	13·1	2·6	41·9	37·2	1·4	0·9	2·0	0·9
		1	22·1	3·2	40·6	32·2	0·5	0·6	0·6	0·3
		2	1·5	0·2	40·2	58·1	—	—	—	—
		3	15·6	4·4	45·1	21·4	3·7	2·0	5·4	2·4
1976	G	TG	13·4	2·1	41·9	37·0	1·1	0·8	2·0	1·0
		1	19·7	2·8	41·5	34·9	—	0·4	0·6	0·2
		2	2·3	—	44·7	33·0	—	—	—	—
		3	18·1	3·5	39·4	25·5	3·2	2·1	5·4	2·7
1977	G	TG	13·4	2·3	41·7	38·2	1·1	0·8	1·8	0·7
		1	25·6	3·4	38·1	31·9	0·2	0·3	0·4	0·1
		2	4·3	0·9	38·6	56·2	—	—	—	—
		3	10·3	2·5	48·6	26·5	3·2	2·0	4·9	2·1
1975	S	TG	11·7	1·8	37·5	43·1	1·2	1·2	2·3	1·1
		1	20·7	2·3	36·6	38·2	0·2	0·8	0·8	0·4
		2	1·0	0·1	31·3	67·6	—	—	—	—
		3	13·5	3·1	44·7	23·7	3·4	2·9	5·9	2·9
1976	S	TG	12·4	1·6	37·9	43·0	1·2	1·1	2·0	0·8
		1	22·6	2·8	34·4	38·8	0·5	0·5	0·4	—
		2	2·9	0·6	31·6	64·2	0·1	0·2	0·3	0·1
		3	11·5	1·4	47·8	25·9	3·1	2·7	5·4	2·2
1977	S	TG	11·8	1·5	36·3	44·3	1·0	1·3	2·5	1·3
		1	22·5	2·0	34·1	39·9	—	0·7	0·6	0·2
		2	2·5	0·4	29·4	67·7	—	—	—	—
		3	10·5	2·1	45·5	25·3	3·0	3·3	6·8	3·6

Sanders, T. H. (1982). *J. Am. Oil Chem. Soc.*, **59**, 348.
G = Gainsville, Florida, USA; S = Stephenville, Texas, USA.
Stereospecific analysis allows fatty acid composition of positions 1, 2 and 3 for peanuts grown in two locations. Fatty acid concentrations as influenced by growing location have a significant influence on peanut triacylglycerol structure.

TABLE 7.26
(a) Variation of Chemical Characteristics of Sunflower Seed with Planting Date

Characteristics and climatic conditions	Planting dates											
	1980					1981						
	3–23 (140)[a]	4–28 (110)	6–2 (108)	6–17 (106)	7–3 (82)	7–18 (105)	3–23 (137)	4–27 (110)	6–2 (109)	6–18 (93)	7–7 (100)	7–23 (102)
Wax in oil (ppm)	645	687	677	1183	2139	673	765	630	649	1225	1136	861
Hull (%)	27.6	25.6	24.9	29.6	32.9	27.2	24.0	24.8	25.5	25.9	23.4	23.6
Oil (%)	44.1	49.2	34.4	40.0	25.3	44.4	47.6	44.7	44.6	42.8	37.1	39.4
Linoleic acid (%)	40.0	52.8	47.2	58.7	65.2	71.1	42.7	47.8	59.4	65.1	74.2	73.6
Hull wax (ppm)	966	747	969	1020	1184	1198	1299	1090	1216	1073	1514	1420
Rainfall[b]	2.53	3.53	4.80	4.75	1.30	7.52	29.69	26.19	32.66	15.37	16.71	8.84
Temperature[c]	35.0	34.1	32.4	32.9	31.3	26.9	31.1	30.9	28.3	26.6	26.1	26.4

[a] Number of days from planting to harvest.
[b] Total rainfall (cm) from 50% flowering to harvest.
[c] Mean high temperature (°C) from 50% flowering to maturity.

(continued)

TABLE 7.26—contd.
(b) Means of the Chemical and Physical Characteristics of Sunflower Seed[a]

Hybrid	Planting year	Irrigation[b] treatment	Hull (%)	Oil (%)	Linoleic acid (%)	Hull (ppm)	Oil (ppm)	Rainfall (cm)	Temp. (°C)
Bushland, (TX)									
MF 894	80	I	27·6	41·3	55·8	1014	999	4·29	32·1
MF 894	81	I	24·5	43·0	60·5	1218	864	21·56	28·2
Florence, SC									
MF 700	82	I	26·8	39·6	55·0	1197	1140	14·25	25·3
		N	27·4	36·3	58·1	1332	1214		
Hysun 101		I	26·9	36·4	58·1	1331	1408		
		N	27·8	35·3	57·0	1530	1715		
DO 842		I	27·5	39·4	56·8	1216	1084		
		N	28·2	35·9	56·6	1068	1116		

[a] Average value over all dates of planting.
[b] I = irrigated; N = non-irrigated.

Morrison, W. H., Sojka, R. E. and Unger, P. W. (1984). *JAOCS*, **61**, 1242.

Seed from sunflower grown at Bushland, Texas in 1980 and 1981 and at Florence, South Carolina were evaluated for the influence of date of planting and irrigation on wax production. Wax content of the hull and the oil was found to be 10% lower in samples from irrigated than for non-irrigated sunflowers. However none of the sunflowers planted on different dates were under severe moisture stress from bloom to harvest. Although the highest positive correlation was found between the linoleic acid content of the oil and the wax content of the hull, those factors which favoured high oil percentage also favoured low wax content.

TABLE 7.27

(a) Effects of Irrigation and Planting Date on Chemical and Physical Composition of Sunflower Seed

Hybrids	Irrigation[b] treatment	Planting dates for 1982														
		3-12 (102)[a]			4-6 (108)			5-1 (86)			8-17 (116)			8-26 (109)		
		Hull (%)	Oil (%)	18:3 (%)	Hull (%)	Oil (%)	18:3 (%)	Hull (%)	Oil (%)	18:3 (%)	Hull (%)	Oil (%)	18:3 (%)	Hull (%)	Oil (%)	18:3 (%)
MF 700	I	27·0	40·0	43·5	23·7	43·2	44·7	30·2	34·1	47·2	26·5	46·9	68·9	26·8	34·0	70·8
	N	29·3	39·0	47·3	26·3	41·2	49·7	30·6	32·4	46·9	26·5	32·8	72·6	26·0	34·8	69·9
HySun 101	I	27·3	42·3	41·7	26·0	42·2	47·3	31·8	30·0	48·1	26·1	33·4	69·8	23·5	34·1	70·4
	N	28·2	39·3	45·6	26·1	40·9	44·3	30·9	33·9	46·1	26·3	31·7	68·8	29·2	31·0	68·2
DO 844	I	24·3	41·4	45·9	26·0	45·2	42·2	29·2	35·8	44·4	28·7	43·6	68·1	25·8	35·0	73·5
	N	26·1	39·7	42·6	28·0	42·5	41·8	29·7	35·4	40·7	28·9	29·1	76·1	26·9	34·0	67·3

Hybrids	Irrigation[b] treatment	8-30 (108)		
		Hull (%)	Oil (%)	18:3 (%)
MF 700	I	—	—	—
	N	25·9	37·4	62·0
HySun 101	I	26·7	35·4	71·0
	N	25·8	34·7	69·1
DO 844	I	30·8	35·2	66·7
	N	26·6	34·9	71·2

[a]Number of days from planting to harvest.
[b]I = irrigated; N = non-irrigated.

(b) Effects of Irrigation and Planting Date on the Wax Content of Sunflower Seed Hull and Oil

Hybrids	Irrigation[a] treatment	Planting dates for 1982									
		3-12		4-6		5-1		8-17		8-26	
		Hull[b]	Oil[b]	Hull	Oil	Hull	Oil	Hull	Oil	Hull	Oil
MF 700	I	813	779	690	684	1109	1436	1689	1313	1683	1491
	N	777	825	910	843	1175	1364	1787	1460	1849	1522
HySun 101	I	648	735	860	860	1573	1916	1436	1487	1868	1924
	N	670	788	825	1753	942	2220	1893	1654	1848	1999
DO 844	I	628	711	923	567	1355	1197	1258	1394	1590	1329
	N	680	656	742	713	999	941	1239	1786	1140	1191

Hybrids	Irrigation[a] treatment	8-30	
		Hull	Oil
MF 700	I	—	—
	N	1495	1273
HySun 101	I	1602	1523
	N	2999	1877
DO 844	I	1539	1303
	N	1576	1411

[a]I = irrigated; N = non-irrigated.
[b]Wax content in ppm.

Morrison, W. H., Sojka, R. E. and Unger, P. W. (1984). *JAOCS*, **61**, 1242.

TABLE 7.28
VARIATION IN FATTY ACID COMPOSITION OF PALM TREES
(a) Individual Palms from One Breeding Plot (Malaysia)

Fatty acids	12:0	14:0	16:0	18:0	18:1	18:2	20:0	18:3
Palm 1	0·05	1·3	40·3	9·1	38·8	9·8	0·4	0·2
Palm 2	0·03	1·0	37·8	6·1	42·2	12·2	0·2	0·3
Palm 3	0·05	1·5	43·9	4·8	33·6	15·4	0·3	0·3

(b) Individual Palms from One Breeding Plot (Malaysia)

Triglycerides	46	48	50	52	54	56
Palm 1	0·6	6·5	38·9	42·1	11·4	0·4
Palm 2	0·4	5·2	34·6	45·3	14·0	0·4
Palm 3	0·9	8·1	43·1	39·3	8·1	0·4

(c) Geographical Variation

Fatty acids	12:0	14:0	16:0	18:0	18:1	18:2	20:0	18:3
Cameroon	0·08	0·9	39·6	5·9	42·0	10·9	0·3	0·4
Malaysia	0·2	1·1	44·2	4·6	39·2	10·0	0·2	0·4

Triglycerides	46	48	50	52	54	56
Cameroon	1·2	4·5	38·4	43·0	12·6	0·2
Malaysia	0·7	7·2	40·0	40·0	11·2	0·7

Jacobsberg, B. (1983). *PORIM Occasional Paper*, October.

TABLE 7.28—contd.

(d)

	1977–78 MARDI		1979–80 PORIM	
	Range	Mean	Range	Mean
12:0	0–0·4	0·1	0·1–1·0	0·2
14:0	0·6–1·7	1·0	0·9–1·5	1·1
16:0	41·1–47·0	43·7	41·8–46·8	44·0
16:1	0–0·6	0·1	0·1–0·3	0·1
18:0	3·7–5·6	4·4	4·2–5·1	4·5
18:1	38·2–43·5	39·9	37·3–40·8	39·2
18:2	6·6–11·9	10·3	9·1–11·0	10·1
18:3	0–0·5	—	0–0·6	0·4
20:0	0–0·8	0·3	0·2–0·7	0·4

Tan, B. K. and Oh, F. C. H. (1981). *PORIM Technology*, May no. 3.

TABLE 7.29
FATTY ACID COMPOSITION OF PALM VARIETIES

	E. Melanococca	Hybrid	E. Guineensis
12:0			0·1
14:0	0·2	0·3	1·0
16:0	23·6	31·7	46·9
18:0	1·0	2·5	5·1
16:1	1·4	0·3	—
18:1	56·9	55·4	34·2
18:2	16·3	9·5	12·3
18:3	0·6	0·3	0·2

Faulkner, H. (1976). Ph.D. thesis, Jan. 1976, Université d'Aix Marseille.

TABLE 7.30
FAT CONTENT AND FATTY ACID COMPOSITION OF THE CLONES OF *Durio zibethinus*

Durio zibethinus clones	Fat content (%)	14:0	15:0	16:0	16:1	17:0	18:0	18:1	18:2	18:3
D_{24}	5·1	0·5	t	39·8	8·5	t	0·8	45·8	1·8	2·7
D_2	5·2	0·5	t	35·9	5·2	t	1·0	51·0	2·4	4·1
D_{66}	3·8	0·8	t	33·8	6·4	t	0·9	48·7	2·8	7·0
D_8	4·2	0·5	t	32·3	2·3	t	2·2	53·6	3·2	6·0

Berry, S. V. (1978). *JAOCS*, **58**, 716.

TABLE 7.31
FATTY ACID COMPOSITION OF TREE NUTS GROWN IN DIFFERENT GEOGRAPHICAL LOCATIONS

		14:0	16:0	16:1	17:0	17:1	18:0	18:1	18:2	18:3	20:0	20:1	22:0 + 22:1
Almond	California I	0·05	6·6	0·35	0·05	0·10	1·20	68·0	23·3	0·05	0·05	0·05	0·20
	California II	0·05	6·5	0·30	0·05	0·10	1·25	67·5	23·9	0·05	0·05	0·05	0·20
	California III	0·05	6·5	0·30	0·05	0·10	1·45	68·9	22·2	0·10	0·05	0·10	0·20
	California IV	0·05	6·8	0·40	0·05	0·10	1·10	67·3	23·8	0·10	0·05	0·05	0·20
	Sfax	0·05	7·2	0·35	0·05	0·10	2·50	67·4	21·9	0·15	0·10	0·10	0·10
	Valencia	0·05	7·1	0·30	0·05	0·10	2·00	67·7	22·4	0·05	0·05	0·10	0·10
Noisette	Levant I	0·05	5·0	0·10	0·05	0·10	2·10	80·3	11·8	0·15	0·10	0·15	0·10
	Levant II	0·05	5·4	0·15	0·05	0·10	1·60	80·2	11·9	0·20	0·10	0·20	0·05
	Tarragona I	0·05	5·8	0·15	0·05	0·10	1·45	66·5	25·1	0·40	0·10	0·20	0·10
	Tarragona II	0·05	5·8	0·15	0·05	0·10	1·40	70·1	21·8	0·25	0·10	0·15	0·05
	Tarragona III	0·05	5·9	0·15	0·05	0·10	1·20	71·5	20·5	0·20	0·10	0·20	0·05
	France	0·05	5·3	0·10	0·05	0·10	1·25	79·1	13·5	0·20	0·10	0·20	0·05
	Rumania	0·05	6·1	0·10	0·05	0·05	2·25	82·5	8·3	0·20	0·15	0·20	0·05

30 m × 0·25 mm Carbowax 20 M, 180°C.
Rugraff, L., Demanze, Ch. and Karleskind, A. (1983). *Parfums cosmetiques aromes*, **43**, February, 59.

TABLE 7.32
VARIATION OF FAT CONTENT AND FATTY ACID COMPOSITION OF PEANUT WITH MATURITY OF SEEDS

Peanut maturity stage (weeks)	Oil content (%)	16:0	18:0	18:1	18:2	20:0	20:1	22:0	24:0
5	25·3								
6	30·8	16·6	2·3	41·7	28·6	1·3	1·8	5·9	1·6
7	34·4	14·4	2·0	44·8	28·6	1·4	1·9	5·2	1·5
8	42·8	13·0	1·9	45·9	31·1	1·2	1·7	3·5	1·6
9	45·6	12·4	2·0	47·7	31·1	1·2	1·4	2·6	1·6
10	46·7	12·4	2·0	49·9	29·3	1·1	1·3	2·4	1·7
11	48·4	12·6	2·0	50·3	29·2	1·1	1·2	2·3	1·3
12	48·2	13·2	2·3	50·7	28·3	1·3	1·1	2·1	1·1

10% DEGS on Chromosorb W-AN-DMCS.
Abdel-Hamid, Y. and Abdel-Rahman, J. (1982). *Fette Seifen Anstrichmittel*, **59**, 285.

7.4 COMPOSITIONS FOR SOME MINOR SEED OILS

TABLE 7.33
FATTY ACID COMPOSITION OF *Linum* SEED OILS

Species[a]	Fatty acid composition (%)						
	Palmitic 16:0	Stearic 18:0	Oleic 18:1	Linoleic 18:2	Linolenic 18:3	Ricinoleic	Unknown[b]
Section Linum							
L. usitatissimum (7)[a]	9.3	2.1	17.2	19.1	52.2	—	++
L. angustifolium (4)	11.1	3.5	17.9	14.5	53.0	—	++
L. bienne (11)	11.6	4.4	16.9	14.7	52.5	—	++
L. grandiflorum var *rubrum* (10)	9.7	3.8	21.5	18.6	46.4	—	++
L. marginale (41)	6.5	2.0	15.5	19.0	57.1	—	—
L. perenne (6)	7.5	2.2	22.5	28.1	39.8	—	++
L. alpinum (6)	7.7	2.3	20.5	27.9	41.7	—	++
L. extraaxillare (1)	7.3	2.0	12.1	28.0	50.7	—	+++
L. anglicum (3)	7.0	2.4	14.3	26.0	50.5	—	+++
L. austriacum (13)	7.7	3.0	21.5	28.6	39.2	—	+++
L. leonii (2)	6.4	1.9	24.3	43.7	23.8	—	+++
L. lewisii (1)	7.7	2.3	20.1	25.4	44.4	—	+++
L. altaicum (1)	8.6	2.4	22.6	24.3	42.2	—	+++
L. mexicanum (1)	8.7	2.3	20.7	28.1	40.3	—	+++
L. narbonense (1)	6.6	1.8	22.0	32.2	37.5	—	+++
Section Dasylinum							
L. hirsutum (2)	6.6	1.8	19.4	27.4	45.0	—	++
L. viscosum (2)	7.2	1.1	13.4	28.2	50.2	—	++
Section Cathartolinum							
L. catharticum (5)	7.6	2.8	13.2	64.3	12.1	—	—

(*continued*)

TABLE 7.33—contd.

Species[a]	Fatty acid composition (%)						
	Palmitic 16:0	Stearic 18:0	Oleic 18:1	Linoleic 18:2	Linolenic 18:3	Ricinoleic	Unknown[b]
Section Linastrum							
L. maritimum (3)	11·0	3·0	13·8	46·1	26·1	—	—
L. strictum (3)	8·9	3·1	7·6	52·9	27·6	—	—
L. rigidum (1)	7·7	1·3	8·1	62·1	20·8	—	—
L. sulcatum (1)	7·9	2·5	12·5	68·7	8·5	—	—
L. imbricatum (1)	8·9	2·5	6·6	75·4	6·5	—	—
L. lundelli (1)	8·4	2·4	8·9	74·6	5·7	—	—
L. tenuifolium (6)	5·0	2·1	8·0	81·5	3·6	—	—
L. salsoloides (3)	5·9	2·7	9·6	78·5	3·4	—	—
Section Syllinum							
L. flavum (8)	7·7	3·7	23·8	47·6	12·6	4·5	—
L. arboreum (2)	6·5	2·9	23·1	50·9	13·6	3·1	—
L. dolomiticum (1)	5·9	2·8	17·8	53·3	16·5	3·6	—
L. campanulatum (2)	5·3	2·3	21·8	51·2	16·4	3·2	—
L. mucronatum (2)	7·2	3·0	20·9	60·8	3·2	5·1	—

Symbols: — = absent, + = less than 5% of total peak area, + + = greater than 5% of total peak area.
[a]Number in parentheses indicates number of accessions analysed for each species.
[b]Unidentified fatty acid methyl ester with ECL of 27·2.
Green, A. G. (1984). *JAOCS*, **61**, 939.

Thirty one *Linum* species representing each of the 5 taxonomic sections of the genus were analysed for their fatty acid composition. Linolenic acid was the major component of the species from the sections *Linum* and *Dasylinum* whereas linoleic acid predominated in those species from the sections *Syllinum*, *Linastrum* and *Cathartolinum*. Ricinoleic acid was only present in the section Syllinum. One unidentified acid was found in species from the sections *Linum* and *Dasylinum*.

TABLE 7.34
FATTY ACID COMPOSITION OF APOCYANACEAE, TILIACEAE CAPPARIDACEAE AND CYPERACEAE

	Family	Oil content (%)	Iodine value	Sap. value	14:0	16:0	16:1	18:0	18:1	18:2
Lochnera pusilla	Apocyanaceae	16·6	75	182	—	25·8	—	7·3	58·6	8·2
Corchosus acutangulus	Tiliaceae	10·0	60	177	1·5	60·2	3·4	4·1	5·9	24·8
Corchosus capsularis	Tiliaceae	12·0	130	193	—	21·4	—	1·9	8·6	67·9
Gynandropsis pentaphylla	Capparidaceae	16·0	121	186	—	16·6	—	9·6	20·0	53·6
Cleome viscosa	Capparidaceae	11·0	80	192	11·9	27·9	—	8·6	22·4	29·2
Cyperus rotundus	Cyperaceae	6·6	69	191	—	43·7	0·4	2·4	34·3	18·9
Cyperus iria	Cyperaceae	10·0	118	182	—	16·6	—	1·7	35·2	46·5

15% DEGS on Chromosorb W (15–60 mesh) and 2% SE30.
Ahmad, F., Ahmad, M. U., Ahmad, I., Ansari, A. A. and Osman, S. M. (1978). *Fette Seifen Anstrichmittel*, **80**, 190.
The seed oils from seven herbaceous species comprising the unusual families Apocynaceae, Tiliaceae, Capparidaceae and Cyperaceae have been analysed by GLC. TLC was also used to determine the presence or absence of oxygenated fatty acids. This is the first report of myristic acid (11·9%) being found in a *Cleome* species. Protein analysis values are also reported.

TABLE 7.35
FAT CONTENT AND FATTY ACID COMPOSITION OF MINOR SEEDS

	Fat Content (%)	Composition (%)						
		12:0	14:0	16:0	18:0	20:0	18:1	18:2
Theobroma cacao	50–55			25·2	35·5		35·2	3·2
Madhuca butyracea	60–65			65·6	3·1		27·4	3·8
Madhuca latifolia	50		0·2	23·5	21·6		39·3	15·4
Madhuca longifolia	50			28·2	14·1		48·8	8·9
Butyrospermum parkii	45–55			5·7	41·0		49·0	4·3
Mimusops njave	—			3·7	35·4	2·1	57·4	1·4
Mimusops heckelii	—			4·2	35·5	1·1	58·5	t
Palaquium oblongifolium	50–55			5·9	54·0		39·9	
Garcinia indica	48		0·4	1·4	60·4		37·8	
Garcinia morella	30			0·7	46·4	2·5	49·5	0·9
Platonia insignis (i)	50		1·0	55·1	6·4	0·3	31·7	2·3
Pentadesma butyracea	32–40			5·4	46·1		48·5	
Allanblackia stuhlmannii	62–67			3·1	52·6		44·1	0·2
Allanblackia floribunda	—			2·9	57·1	0·2	39·4	0·4
Allanblackia parviflora	—		1·5	2·3	52·0	0·3	43·9	
Gnetum scandens	14			12·0	54·7		30·3	3·3
Shorea stenoptera	—			18·0	43·3	1·1	37·4	0·2
Shorea robusta	20			8·3	34·7	12·3	41·9	2·8
Vateria indica (ii)	22–27			9·7	40·7	4·6	42·2	2·3
Rhus succedanea (iii)	65		1·9	67·5	11·6		13·6	
Mangifera indica (iv)	6–12			8·0	42·6	1·6	42·0	3·4
Cinnamomum camphora	42	95·0					5·0	
Actinodaphne hookerii	48	96·0					4·0	
Actinodaphne angustifolia	37	90·0					10·0	
Litsea longifolia	29	88·3		3·4	2·4		5·9	
Litsea cubeba	22	96·1					2·2	
Litsea zeylanica (v)	36	86·0	4·0				4·0	3·0

Species									
Litsea sebifera	35	96.3						2.3	18.2
Laurus nobilis	24–30	43.1						32.5	t
Salvadora oleoides	40	35.6						8.3	
Salvadora persica (vi)	—	19.6						5.4	
Cocos nucifera	50	50.0		6.2	1.0			16.5	1.0
Elaeis guineensis	44–53	0.6		4.5				40.6	6.2
Myristica fragrans (vii)	38–43	71.8		19.5	2.8			5.2	1.5
Myristica malabarica	41	39.2	0.4	6.5	1.2			44.1	1.0
Virola otoba	67	73.1		49.9	2.4			5.5	
Virola surinamensis (viii)	65	73.2		14.3				6.3	
Virola bicuhyba	—	66.6		13.3				6.6	3.0
Pycnanthus kombo (ix)	—	61.6	5.5	3.6	1.6			5.7	
Bombax malabarica	18–26			28.3	7.3			49.9	14.5
Sapium sebiferum	55–78	0.3		62.3	5.9			27.4	
Canarium commune (x)	68		4.2	30.5	10.2			39.9	18.7
Canarium ovatum	78			38.2	1.8			59.8	
Dacryodes rostrata	32			12.7	30.9			49.5	2.8
Nephelium lappaceum (xi)	34–40		1.0	2.0	13.8	3.1		45.3	
Nephelium mutabile	60–72			3.0	31.0	34.0		43.7	
Irvingia gabonensis (xii)	61	58.6		2.0	1.1	22.3		1.8	
Irvingia barteri	69	38.8						10.6	
Irvingia oliveri (xiii)	60	39.0						5.0	
Lophira alata (xiv)	31–40		33.4	28.8	2.6			14.0	11.5
Lophira procera (xv)	—		50.6	37.9	2.2			11.5	26.3
Erisma calcaratum	53	23.9	55.5	18.9	0.9			2.8	
Acrocomia sclerocarpa (xvi)	—	44.9	0.3	7.6	2.6			16.5	1.6
Trichilia emetica (xvii)	55–65		13.4	38.3	2.2			48.5	10.4
Caryocar villosum	72			48.4	0.9			46.0	3.3
Myrica cerifera	20		1.4	45.0	0.3			0.6	
Myrica cordifolia	20	0.3	33.0	51.8	42.6			42.0	
Hodgsonia capniocarpa	66		47.0	8.0		1.6			3.7

(continued)

TABLE 7.35—contd.

In addition the following acids are present: (i) palmitoleic 3·2%; (ii) linolenic 0·5%; (iii) sat. C_{22} and C_{23} + dibasic acids 6·0%; (iv) linolenic 1·4%; (v) capric 3·0%; (vi) capric 1·0%; (vii) palmitoleic 4·8%; (viii) capric 0·5%; (ix) myristoleic 23·6%; (x) linolenic 0·7%; (xi) gadoleic 4·2%; (xii) caprylic 3·1%; (xiii) caprylic 0·3%; (xiv) behenic 34·3%, lignoceric 6·8%, C_{22} unsat. 4·3%; (xv) behenic 20·9%, lignoceric 0·5%, C_{22} unsat. 2·2%; (xvi) caprylic 7·8%, capric 5·6%; (xvii) linolenic 1·0%.
Banerji, R., Chowdhury, A. R., Misra, G. and Nigam, S. K. (1984). *Fette Seifen Anstrichmittel*, **86**, 279.

M. butyracea yields Phulwara butter which melts at 39–47°C.
M. latifolia and *M. longifolia* yields Mowrah butter (illipe butter).
B. parkii yields shea butter also called Karite or Galam butter.
M. njave yields Baku butter which is used for cooking.
M. heckelii yields Dumori butter which is edible.
P. oblongifolium yields 50–55% Njatus tallow.
G. indica yields a brittle Kokum butter m.p. 40–43°C.
G. morella yields a fat known as Gamboge or Gurgi butter.
P. insignis yields Bacury butter m.p. 54–56°C.
P. butyracea yields Lamy butter with a melting point of 28–37°C.
V. indica yields Malabar tallow m.p. 30–40°C.
M. indica yields the solid edible Mango butter.
C. camphora is a native of China, Japan and Formosa which yields a solid fat, Kusu, rich in lauric acid. Litsea fats do not appear to have attained much practical importance.
S. oleoides yields a hard yellow solid, Khakan fat m.p. 35°C.
Several species of *Acrocomia* grow in tropical America.
A. sclerocarpa yields macaja butter.
M. malabarica is called the false nutmeg or Bombay Mace tree and yields a fat melting at 32°C which is used as an embrocation in rheumatism.
V. otoba yields otaba butter melting at 34°C.
P. kombo yields a red brown, aromatic fat called Kombo fat.
C. commune yields java almond fat.
N. lappaceum yields Rambutan tallow which is unlikely to attain economic importance because the seeds are available for a short period only.
L. alata yields Niam fat characterised by a high proportion of behenic acid.
Myrica species yield Bayberry or myrtle wax used in fancy candles.

TABLE 7.36
OIL CONTENT, IODINE VALUE AND FATTY ACID COMPOSITION OF MINOR SEEDS

Variety	Family	Oil content	Iodine value	12:0	14:0	16:0	18:0	20:0	22:0	18:1	18:2	18:3
Acacia fornesiana	Leguminosae	2.0	70.5	3.5	4.2	1.9	4.3	3.5	5.1	76.1	0.8	0.6
Amaranthus tricolor	Amarantaceae	5.9	81.8	0.2	0.8	1.9	3.9	0.8	1.4	90.7	0.2	0.1
Bixa orellana	Bixaceae	8.5	52.5	0.3	1.2	1.6	26.9	11.9	7.3	48.0	1.5	2.1
Cassia auriculata	Leguminosae	2.5	96.5	3.5	4.3	3.0	7.0	4.0	2.2	43.4	31.7	0.9
Celosia cristata	Amarantaceae	23.0	72.5	—	0.4	14.7	8.7	2.2	0.8	64.4	8.7	—
Duranta repens	Verbenaceae	3.5	105.0	0.3	0.3	8.2	12.7	1.0	0.7	42.5	29.0	5.3
Holmskioldia sanguinea	Verbenaceae	9.5	110.3	0.5	0.4	6.3	10.5	1.5	0.5	43.0	32.9	4.4
Hymenantherum tenuifolium	Compositae	35.5	117.2	0.8	1.5	2.0	3.6	1.3	2.2	48.3	40.2	—
Ipomoea involucrata	Convolvulaceae	13.8	48.7	1.3	4.0	15.0	11.5	4.1	2.3	53.8	0.3	1.3
Ipomoea pestigridis	Convolvulaceae	6.9	48.9	2.6	3.6	2.3	11.4	6.7	6.2	58.5	0.1	—
Ipomoea sepiaria	Convolvulaceae	34.7	92.9	2.2	3.8	4.0	6.0	2.2	4.9	53.6	19.5	3.8
Lantana sellowiana	Verbenaceae	7.5	115.0	0.2	0.2	4.8	10.7	1.7	0.6	41.5	35.8	4.5
Physalis maxima	Solanaceae	14.3	83.1	2.8	12.3	5.9	3.5	4.8	6.2	23.6	31.0	4.0
Salvia farinacea	Labiatae	27.4	77.1	2.3	2.3	2.6	2.6	1.1	4.9	82.2	1.9	—
Sida spinosa	Malvaceae	5.6	80.0	0.3	0.4	14.2	3.6	2.5	2.4	67.6	6.0	3.0

Badami, C. and Takkar, J. (1984). *Fette Seifen Anstrichmittel*, **86**, 115.

(*continued*)

TABLE 7.36—contd.

Oleic acid is the major fatty acid in fourteen seed oils from nine plant families. The exception is *Physalis maxima*. Linolenic acid is found in eleven samples and lauric acid is found in all of the oils except the oil from *Celosia cristata*.
Acacia farnesiana is a thorny bush whose flower yields the valuable cassie perfume.
Amaranthus tricolor is cultivated throughout India and Sri Lanka.
Bixa orellana has become naturalised in India and yields the orange red dye 'annatto'.
Cassia auriculata is a shrub which is valuable as a tanning material.
Celosia cristata is cultivated for ornamental purposes.
Duranta repens is an evergreen shrub.
Holmskioldia sanguinea is a climbing shrub which grows in gardens throughout India.
Hymenantherum tenuifolium can be propagated by seeds, division of roots or by cuttings.
Ipomoea involucrata is a member of a large genus containing twining, creeping, floating or erect herbs.
Lantana sellowiana is a low, erect or sub-scandant shrub.
Physalis maxima is a member of a genus of herbaceous annuals.
Salvia farinacea is a useful bedding plant in Deccan gardens.
Sidi spinosa is grown in hotter parts of India and Sri Lanka.

TABLE 7.37
OIL CONTENT AND FATTY ACID COMPOSITION OF KERNEL OILS

	Oil content	14:0	16:0	18:0	18:1	18:2	20:0	20:1	22:0	24:0
Pentaclethra macrophylla	45·9	0·2	3·7	2·3	31·3	40·4	2·5	2·3	8·5	8·8
Allanblackia floribunda	67·6		0·8	55·9	43·3					
Panda oleosa	50·5	1·5	32·0	7·1	30·2	29·2				
Treculia africana[a]	11·8		25·7	14·2	32·7	25·8				
Desplatzia dewevrei	20·4	0·2	37·8	7·4	18·1	35·0	1·5			
Garcinia kola[b]	2·1	1·3	19·0	13·1	38·4	23·7				
Milletia laurentii	22·9	0·1	10·5	2·9	44·9	17·6	1·8	9·9	10·1	2·2

In addition, species (*a*) has 1·6% $C_{16:1}$ and species (*b*) has 1·3% $C_{12:0}$ and 3·2% $C_{18:3}$.
10% DEGS on Chromosorb W (80–100 mesh).
Foma, M. and Abdala, T. (1985). *JAOCS*, **62**, 910.
Kernels of some wild plants from Zaire have been examined for their oil contents and fatty acid compositions.

TABLE 7.38
OIL CONTENT AND FATTY ACID COMPOSITION OF CYCLOPROPANE CONTAINING SEED OILS.

	Oil content (%)	12:0	14:0	16:0	18:0	18:1	18:2	18:3	20:0	A	B	C	D
Hibiscus surathensis	16.6		0.2	20.0	3.2	24.8	43.9		0.7	3.7	0.7	2.1	0.5
Hibiscus vitifolius	13.3		0.7	30.1	4.3	15.2	44.8		0.7	3.0	0.6	0.5	0.1
Hibiscus hirtus	14.6		0.3	15.1	3.0	8.8	67.6		0.7	2.0	1.1	0.5	t
Hibiscus punctatus	13.0	0.8	0.2	17.0	1.9	14.1	53.3			8.4	1.8	0.3	t
Hibiscus zeylanicus	13.6	0.8		26.2	3.3	5.9	56.3	2.1	0.3	4.0	3.9	t	t
Hibiscus micranthus	15.2		0.5	18.6	3.5	10.1	59.8		1.0	1.7	3.1	1.0	0.5
Hibiscus solandra	15.7		0.5	17.3	3.9	8.8	64.4		0.8	1.7	0.7	1.3	0.5

A = malvalic acid; B = sterculic acid; C = dihydrosterculic acid; D = epoxy acids.
Sundar Rao, K. and Lakshminarayana, G. (1985). *JAOCS*, **62**, 714.
As part of a study to augment oil resources, 7 *Hibiscus* species were studied and found to contain malvalic, sterculic and dihydrosterculic acids in low percentages.

TABLE 7.39
FATTY ACID COMPOSITIONS OF SPICES

			16:0	16:1	16:2	18:0	n-12 18:1	n-9 18:1	n-7 18:1	n-6 18:2	n-12 18:2	18:3
Asclepias syriaca	Asclepiaceae		4·9	4·7	1·2	2·1		16·5	12·3	54·9		1·5
Asclepias incarnata	Asclepiaceae		5·8	7·5	0·8	1·9		19·6	12·8	48·6		1·3
Vincetoxicum nigium	Asclepiaceae		4·2	10·6	1·0	2·4		15·7	6·2	57·0		1·0
Leucas cephalotes	Lamiaceae		12·4			4·5	4·3	37·0		22·2	16·5	
Anethium graveolens	Apiaceae	Dill	4·5			1·4	72·5	7·6		12·2		
Carum carvi I	Apiaceae	Kummel	4·5			1·3	43·6	12·1		36·1		
Carum carvi II	Apiaceae		5·0			1·3	42·5	15·2		34·0		
Coriandrum sativum	Apiaceae	Coriander	3·7			1·1	73·2	7·2		12·9		
Cuminum cyminum	Apiaceae		4·1			0·9	52·7	8·8		30·4		
Foeniculum vulgare	Apiaceae	Fennel	3·8			0·9	76·6	4·5		12·1		
Petroselinum sativum	Apiaceae		3·9			0·9	79·9	3·8		10·8		
Pimpinella anisum	Apiaceae	Anis	4·0			1·3	55·8	15·3		20·6		

50 m WCOT Capillary Silar 5CP and HPLC on Li Chrosorb RP18.
Seher, A. and Fiebig, H. J. (1983). *Fette Seifen Anstrichmittel*, **85**, 333.
Triglyceride structures of these fats have been determined with the help of HPLC.

TABLE 7.40
FATTY ACID COMPOSITION OF SEVERAL *Bauhinia* SPECIES

	14:0	15:0	16:0	16:1	17:0	18:0	18:1	18:2	20:0	20:1
Bauhinia retusa	0·2	0·1	22·0			10·7	28·3	34·2	4·6	
Bauhinia variegata	0·3		22·7	2·2		17·9	14·1	43·4	1·6	
Bauhinia triandra	0·2		23·7			10·1	18·8	44·9	2·2	
Bauhinia acuminata[a]	0·1		12·5			9·5	26·3	49·9	1·0	0·9
Bauhinia tomentosa	0·3		15·4			8·5	23·9	51·9		
Bauhinia racemosa[b]	0·5	0·9	18·7		1·0	11·6	23·9	36·5	1·5	

[a] Also includes 0·1% C_{12}.
[b] Also includes 1·8% C_8, 1·3% C_{10}, 1·5% C_{12}, 0·9% C_{13}.
10% DEGS on Chromosorb W.
Chowdhury, Ar. R., Banerji, R., Misra, G. and Nigam, S. K. (1984). *Fette Seifen Anstrichmittel*, **86**, 237. Mineral compositions of the seeds of these species of *Bauhinia* are also reported.

TABLE 7.41
OIL CONTENT, IODINE VALUE AND FATTY ACID COMPOSITION OF *Convolvulaceae* GENERA

	Oil content	Iodine value	16:0	18:0	18:1	18:2	18:3	20:0	22:0
Convolvulus arvensis	9·0	119	7·6	12·1	22·4	40·6	9·2	5·3	2·8
Evolvulus alsinoides	8·4	109	8·4	14·8	23·8	37·8	6·5	4·4	4·3
Ipomoea biloba	15·0	110	10·0	12·0	21·6	41·3	6·0	5·5	3·6
Ipomoea hederacea	14·7	97	9·4	19·6	24·0	27·8	8·9	6·4	3·9
Ipomoea quamoclit	16·5	110	7·3	13·1	30·0	34·8	7·4	4·4	3·0
Ipomoea tuberosa	6·7	110	6·6	15·8	25·9	35·7	7·9	3·3	4·8

Minor Seed Oils Convolvulaceae

Badami, R. C. and Thakkar, J. (1984). *Fette Seifen Anstrichmittel*, **86**, 203.
Minor seed oils have been examined and no unusual acids have been detected.

TABLE 7.42

OIL CONTENT, IODINE VALUE, SAPONIFICATION EQUIVALENT AND FATTY ACID COMPOSITION OF MINOR SEED OILS

Minor seed oils	Oil content (%)	Iodine value	Sap. equiv.	8:0	10:0	12:0	14:0	16:0	18:0	18:1	18:2	18:3	20:0	22:0
Androgaphis paniculata	39.4	108.8	273.3	—	—	1.4	2.2	25.2	5.4	4.8	58.9	—	1.1	1.0
Calliopsis elegans	7.2	104.2	273.9	—	—	2.8	6.2	11.1	4.7	21.2	48.3	—	2.0	3.7
Crotalaria heyneana	3.4	113.9	272.5	—	1.3	0.8	1.5	16.8	8.0	7.4	60.5	—	1.6	2.1
Corochorus trilocularis	8.1	82.9	267.7	—	—	8.7	2.1	22.0	7.0	12.1	36.6	3.6	4.7	2.2
Emblica officinalis	22.4	84.4	270.1	—	—	5.7	3.8	12.4	10.7	29.8	23.8	5.9	5.2	2.7
Mangifera indica	12.0	45.6	277.4	—	—	—	0.8	12.7	37.4	38.8	6.8	—	2.2	1.3
Pergularia daemia	9.4	62.7	266.9	0.5	1.0	0.3	1.2	43.8	6.9	14.6	29.3	—	1.2	1.2
Sopubia dulphinifolia	5.2	103.2	273.3	1.3	1.2	1.0	1.6	6.1	8.7	31.7	42.7	—	2.8	2.9

Badami, R. C. and Alagawadi, K. R. (1983). *Fette Seifen Anstrichmittel*, **85**, 197.
Eight minor seed oils have been analysed. The seeds from the shrub *Androgaphis paniculata* contain 39.4% oil whereas the seeds of the trees *Emblica Officinalis* and *Mangifera indica* contain 22.4% and 12.0% oil respectively. *Pergularia daemia* is rich in stearic acid (37.4%) whilst *Crotalaria heyneana*, *Androgaphis paniculata* and *Calliopsis elegans* seed oils are rich in linoleic acid (60.5%, 58.9% and 48.3% respectively). In contrast the seed oils from *Corochorus trilocularis* and *Emblica officinalis* contain 3.6% and 5.9% linolenic acid. The iodine values of 63–85 for the oils of *Pergularia daemia*, *Corochorus trilocularis* and *Emblica officinalis* suggest that they could be useful as non-drying oils.

TABLE 7.43
OIL CONTENT AND FATTY ACID COMPOSITION OF SPECIES OF GENERA Stizolobium, Mucuna AND Lucuma

	Oil content (%)	12:0	12:1	14:0	16:0	16:1	18:0	18:1	18:2	18:3	20:0	22:0
Stizolobium aterrimum Piper and Tracy	4.3		0.5	0.2	30.2	0.6	8.2	10.5	42.6	7.4	tr	
Mucuna pruriens D.C. (6) (S. pruriens Pers.)	3.3				26.5	0.9	6.3	11.4	42.1	7.0	1.5	3.6
(6)					23.8	1.0	5.6	10.9	46.6	9.2	0.9	2.0
(7)	4.3			1.3	53.7		19.8	18.1	10.2			
Mucuna flagellipes T. Vogel (8)	3.7				23.0	tr	6.8	30.9	37.5	1.8	tr	
Mucuna imbricata (9)	7.0				15.6		4.7	37.5	32.4			
Lucuma caimito Roem.	13.2	tr	tr	0.4	23.1	tr	8.8	57.2	9.7	0.3		
Calocarpum mammosum Pierre (17) (Lucuma mammosa Gaertn)	57.0				9.4		20.9	52.1	12.8		0.02	
(18)							37.5	38.2	24.2			
Lucuma sacilifolia (6)		3.0		2.5	16.1		9.2	50.1	18.7	0.7	1.7	

2.1 m × 0.32 mm column of 17% DEGS on Chromosorb W (80–100 mesh) at 190°C.
Schuch, R., Baruffaldi, R. and Gioielli L. A. (1984). *JAOCS*, **61**, 1207.
These fatty acid compositions are from Brazilian plants.
The genus *Stizolobium* was formerly included in the genus *Mucuna* but it is now distinguished from the latter by its seeds. The species *S. aterrimum* is the most important in Brazil and its fatty acid composition is compared with literature values for *Stizolobium* and *Mucuna*.

TABLE 7.44
FATTY ACID COMPOSITION OF SEED OILS FROM *Brassicaceae*

Seed oils		12:0	14:0	15:0	16:0	16:1	17:0	17:1	18:0
Brassica chinensis	Chinese cabbage	0·1	0·1	—	2·3	0·2	—	0·1	0·8
Brassica oleracea var capitata	White cabbage	0·1	0·1	t	3·8	0·2	t	t	0·9
Brassica oleracea var sabauda	Savoy	t	0·1	—	3·9	0·3	t	0·1	0·8
Brassica oleracea convar acephala	Kale var sabellica	0·1	0·1	t	3·6	0·3	t	0·1	0·9
Brassica oleracea var capitata	Red Cabbage	t	t	—	3·5	0·2	0·1	0·2	0·8
Brassica oleracea var gemmifera	Brussels sprouts	t	0·1	t	4·1	0·2	0·1	0·1	1·0
Brassica oleracea Convar botrytis	Cauliflower	t	0·1	t	4·0	0·3	0·1	0·1	0·8
Brassica oleracea convar acephala	Kohlrabi var gongylodes	—	0·1	t	3·4	0·3	0·1	0·1	0·8
Brassica napus var napobrassica	Swede	t	0·1	t	3·4	0·2	t	0·2	1·2

Nasirullah, I., Werner, G. and Seher, A. (1984). *Fette Seifen Anstrichmittel*, **86**, 264.

Edible parts and seeds of 15 vegetables from *Brassicaceae, Apiaceae and Asteraceae* have been analysed. Whereas seed oils are rich in erucic acid, this acid is often missing from lipids of edible vegetative organs.

n-9 18:1	n-7 18:1	n-6 18:2	n-3 18:3	20:0	n-9 20:1	n-7 20:1	n-6 20:2	n-3 20:3	22:0	n-9 22:1	n-7 22:1	22:2	24:0	24:1
13·4	1·1	13·4	8·7	0·7	5·9	1·3	0·4	0·1	0·9	46·6	1·2	0·9	0·03	1·5
11·6	1·5	14·7	9·7	0·7	6·7	1·6	0·7	0·1	0·6	43·6	0·8	0·8	0·4	1·4
12·0	1·7	13·3	11·5	0·6	6·9	1·7	0·7	0·1	0·4	42·6	0·8	0·7	0·3	1·5
17·5	1·9	13·9	8·3	0·6	9·5	1·4	0·6	0·2	0·4	38·2	0·7	0·5	0·2	1·0
12·1	1·3	15·8	10·2	0·5	6·3	1·6	0·7	0·1	0·6	42·5	0·9	0·9	0·3	1·4
13·6	1·3	13·4	11·2	0·7	7·8	1·5	0·6	0·2	0·7	40·7	0·7	0·7	0·3	1·0
16·6	2·0	13·7	9·3	0·4	7·8	1·7	0·6	t	0·5	38·7	0·7	0·6	0·3	1·7
10·8	1·4	12·4	11·0	0·5	6·0	1·8	0·7	0·1	0·5	47·0	0·9	0·8	0·2	1·1
19·8	2·4	13·9	12·3	0·6	10·7	1·4	0·5	0·1	0·2	31·1	0·4	0·3	0·1	1·1

TABLE 7.45
FATTY ACID COMPOSITION OF EDIBLE PORTIONS OF *Brassicaceae*

Edible portions of vegetables		10:0	12:0	14:0	15:0	16:0 u	16:0	16:1	16:2
Brassica chinensis	Chinese cabbage	t	0·1	0·4	0·3	0·1	16·1	0·8	0·2
Brassica oleracea var capitata	White	0·1	0·2	0·3	0·4	0·2	14·5	0·9	0·1
Brassica oleracea var Sabauda	Savoy	0·1	0·1	0·3	0·3	0·2	22·1	0·5	0·1
Brassica oleracea Convar acephala var Sabellica	Kale	t	t	0·2	0·3	0·2	13·0	0·4	—
Brassica oleracea var Capitata	Red Cabbage	—	0·1	0·2	0·5	t	18·6	0·6	0·1
Brassica oleracea var gemmifera	Brussels sprouts	0·1	0·2	0·5	0·4	t	14·5	0·5	—
Brassica oleracea convar botrytis	Cauliflower	—	0·1	0·2	0·4	t	17·4	1·7	0·4
Brassica oleracea convar acephala var gongylodes	Kohlrabi	0·1	0·2	0·3	0·4	0·1	20·8	1·6	0·1
Brassica napus var napobrassica	Swede	0·1	0·4	0·4	0·6	0·3	14·5	2·3	0·2
Lactuca sativa var capitata	Cabbage	t	0·1	0·4	0·2	0·1	13·9	1·4	—
Petroselinum crispum Nym spp. crispum	Parsley	t	0·1	0·3	0·2	t	12·8	1·7	—

Nasirullah, I., Werner, G. and Seher, A. (1984). *Fette Seifen Anstrichmittel*, **86**, 264.
The fatty acids of the edible parts of 15 vegetables belonging to the families *Brassicaceae, Apiaceae* and *Asteraceae* are reported.

Varietal Differences in Fatty Acid Compositions

16:3 n-3	17:0 u	17:0	18:0	18:1 n-9	18:1 n-7	18:2 n-6	18:3 n-3	20:0	20:1	20:2 n-6	20:3 n-3	22:0	22:1 n-9	24:0	24:1
3·0	1·3	0·2	1·9	1·3	5·6	7·5	59·5	0·6	0·3	—	—	0·3	—	0·4	0·3
1·1	0·4	0·2	1·1	1·3	5·4	14·8	57·0	0·3	0·3	0·2	0·1	0·4	t	0·2	0·6
0·9	2·2	0·2	1·2	0·7	5·7	15·7	46·4	0·8	0·3	0·3	—	0·7	—	0·5	0·8
6·6	4·4	0·3	1·2	0·7	2·3	18·5	50·0	0·3	0·3	0·1	0·2	0·2	0·3	0·2	0·3
0·4	0·3	0·5	1·9	0·5	2·9	29·9	39·8	1·0	0·6	t	0·1	0·5	—	1·0	0·5
4·1	1·8	0·2	0·9	0·6	3·1	11·4	58·7	1·6	0·4	0·2	0·2	0·1	—	0·2	0·3
0·5	0·6	0·3	1·7	0·8	11·9	12·8	48·9	0·5	0·4	—	0·2	0·2	0·3	0·3	0·4
0·5	0·3	0·3	1·9	4·2	6·0	17·9	41·9	1·3	0·3	—	—	0·4	—	1·1	0·3
0·3	0·4	0·4	0·7	0·6	6·8	20·7	50·8	—	0·2	—	—	—	—	0·2	0·1
0·1	0·1	0·2	0·7	0·6	0·3	23·0	54·6	0·4	0·6	0·2	0·2	0·9	0·1	1·3	0·6
11·8	0·3	0·2	0·6	0·6	0·2	25·6	41·7	0·6	0·3	0·3	0·1	1·1	—	1·4	0·1

TABLE 7.46
OIL CONTENT AND FATTY ACID COMPOSITION OF MINOR SEED OILS

Variety	Oil content	14:0	16:0	16:1	18:0	18:1	18:2	18:3	20:0
Cedrella odorata	18·00	—	12·5	0·5	4·9	11·6	51·4	19·1	—
Lovoa trichilloides	25·86	0·3	12·7	0·2	—	20·7	63·3	0·7	—
Khaya senegalensis	45·50	4·6	11·3	0·1	13·9	59·4	9·7	0·6	0·5
Carapa procera	53·45	0·2	26·8	—	4·1	62·5	6·3	—	0·1
Enthandrophragma angolense[a]	52·00	0·1	7·1	17·7	18·8	40·4	24·3	—	1·7
Azadirachta indica	17·44	—	35·8	—	12·0	36·1	15·1	—	2·0
Terminalia Superba[b]	14·50	3·0	33·7	0·7	5·6	25·8	28·4	0·5	2·5
Terminalia glucausens[c]	17·50	0·1	35·0	0·5	4·8	32·7	26·7	—	—
Terminalia catappa	40·15	1·2	30·2	—	5·8	41·5	19·1	0·9	1·3

In addition species (a) has 0·2% $C_{12:0}$; (b) has 0·6% $C_{12:0}$ and 1·2% $C_{22:0}$; and (c) has 0·4% $C_{10:0}$. Balogun, A. M. and Fetuga, B. L. (1985). *JAOCS*, **62**, 529.

As part of a programme to uncover the potential of 60 underutilised plants in Nigeria, several members have shown promise as alternative sources of proteins and oils.

TABLE 7.47
FATTY ACID COMPOSITION OF PLANTS USED FOR CROSSING WITH SOYABEANS

	16:0	18:0	18:1	18:2	18:3
Centrosema pubescens	10·1	3·4	30·2	24·9	31·5
	11·1	1·2	27·9	23·3	36·5
Centrosema virginionum	7·5	0·8	33·5	19·7	38·5
	13·5	1·6	38·1	20·8	26·1
Centrosema arenarium	11·5	3·2	36·1	36·1	13·3
Centrosema laurifolia	7·8	2·1	74·7	3·9	11·5
Centrosema sp.	12·4	1·7	36·7	20·2	28·9
Clitoria ternatea	16·4	5·2	50·3	26·7	1·4
	15·9	7·8	54·7	20·5	1·2
	17·9	6·0	44·3	29·7	2·0
	17·0	8·5	50·7	22·3	1·6
Teramnus uncinatus	12·5	8·8	11·9	36·7	30·7
	12·0	7·7	8·5	39·1	32·9
	12·2	10·7	10·7	41·3	25·0
	12·1	5·5	10·9	39·5	32·0
Teramnus volubilis	9·4	2·1	18·7	50·1	19·8
Paraglycine radicosa	11·6	4·1	32·7	12·3	39·3
	10·2	6·8	26·2	9·1	47·6
	15·4	2·8	31·8	8·5	41·6
	15·8	5·5	4·6	14·3	59·7
	10·4	4·0	25·9	9·7	50·2
Paraglycine hedyseroides	17·5	4·3	36·9	13·4	28·1

Hammond, E. G. and Fehr, W. R. (1975). *Fette Seifen Anstrichmittel*, **77**, 97.

As part of a survey of *Glycine* species to determine if any have lower linolenic acid contents than *Glycine max.*, it was found that all had higher levels. Further analyses for soybean crosses are given.

TABLE 7.48
OIL CONTENT, UNSAPONIFIABLE CONTENT, IODINE VALUE AND FATTY ACID COMPOSITION OF MINOR SEED OILS

Variety	Family	Oil content	Unsap.	Iodine value	14:0	16:0	18:0	18:1	18:2	18:3	20:0	22:0
Celosia pyramidalis	Amarantaceae	22.5	0.3	75.0	0.6	17.2	6.5	64.4	8.7	—	1.9	0.7
Cryptostegia grandiflora	Asclepiadaceae	9.3	0.9	113.7	—	5.9	4.1	52.3	34.4	1.8		
Gomphrena globosa	Amarantaceae	11.8	0.6	82.6	1.1	18.5	9.5	45.7	23.9	—		
Isotoma longiflora	Campanulaceae	3.9	0.5	84.1	0.2	20.3	10.9	34.7	25.3	2.8		
Jasminium officinale	Oleaceae	2.7	0.3	84.9	1.2	4.4	6.0	80.4	6.9	—		
Jatropha gossypifolia	Euphorbiaceae	18.0	0.4	106.0	—	14.8	5.4	37.9	41.2	—		
Leucas linifolia	Labiateae	8.4	0.7	112.2	—	5.9	5.4	35.2	44.3	—		
Opuntia dillenii	Cactaceae	7.5	0.5	115.1	1.0	18.7	5.9	17.3	56.1	—		
Pavetta indica	Rubiaceae	3.6	0.8	105.4	—	14.3	8.7	33.5	40.3	1.5		
Psychotria dalzellii	Rubiaceae	6.3	0.4	87.7	1.9	26.4	8.7	23.5	38.1	—		
Sida humilis	Malvaceae	4.8	0.9	72.8	0.5	17.0	4.0	65.0	5.7	1.8		
Toddalia asiatica	Rutaceae	5.7	0.7	107.7	—	1.4	19.7	37.5	38.9	1.3		

Badami, R. C. and Thakkati, J. K. (1984), *Fette Seifen Anstrichmittel*, **86**, 165.

Twelve seed oils from ten plant families have been examined of which 9 varieties are reported to be of medicinal use.

Celosia pyramidalis has flowers in feathery plumes.
Cryptostegia grandiflora is a climbing evergreen which grows easily in marshy places. It has toxic leaves and contains rubber.
Gomphrena globosa is an erect, hairy, dichotomously branched annual. It is easily propagated by seeds.
Isotoma longiflora is a herb and is sometimes grown for ornamental purposes. The plant is poisonous when taken internally.
Jasminium officinale is cultivated in gardens and is anthelmintic, diuretic and emmengogue.
Jatropha gossypifolia is a shrub which can be raised from seeds. The seed oil is used in leprosy treatment.
Leucas linifolia is widely distributed throughout the plains in India.
Opuntia dillenii is an introduced plant in India and is used in many herbal medicines.
Pavetta indica is distributed throughout India.
Psychotria dalzellii is a large shrub common in green forests.
Sida humilis is distributed throughout the hotter parts of India.
Toddalia asiatica is distributed in most parts of India and is used as a febge.

TABLE 7.49
CHEMICAL COMPOSITION OF LEGUMINOUS SEEDS

Name of seeds	Oil[a] content (%)	Protein[a] (%)	Fatty acid composition[b]								
			14:0	16:0	16:1	18:0	18:1	18:2	20:0	22:0	24:0
Albizzia lebbek Benth.	5.3	29.5	—	12.0	0.5	2.7	21.4	57.1	2.5	2.9	0.9
A. lucida Benth.	12.7	30.2	0.5	20.6	0.3	4.4	20.3	49.0	2.9	1.4	—
A. richardiana King and Prain	2.8	28.4	—	15.0	—	13.7	30.9	36.2	3.7	0.5	—[c]
Entada phaseoloides Merrill Syn. E. scandens Benth.	8.0	27.7	Tr.	12.5	0.5	8.8	57.5	12.5	0.4	1.1	0.6[d]
Leucaena leucocephala (Lamk.) Wit.	7.5	24.5	0.1	15.8	0.2	6.2	15.9	56.8	2.7	2.1	—
Parkia biglandulosa W. and A.	19.8	28.4	0.2	25.6	7.0	32.7	26.8	2.9	3.9	Tr.	—[e]
Pithecolobium bigemina (L) Mart. Syn. Inga bigemina Hook and Arn.	0.7	21.3	0.5	39.5	—	7.3	24.9	23.6	4.0	—	—
P. dulce Benth.	13.0	37.5	Tr.	12.3	0.3	3.3	51.1	13.3	2.5	10.0	5.3[f]

[a]Percentage by weight.
[b]Percentage by area.
[c]Also contains 17:0, 0.2%.
[d]Also contains 8:0, 0.9; 9:0, 0.2; 10:0, 0.1; 12:0, 0.2; 12:1, 0.1; 13:0, 0.3; 13:1, Tr.; 14:1, 0.1; 15:1, 1.6; 17:0, Tr.; and 20:1, 0.7%.
[e]Also contains 15:1, 0.8%.
[f]Also contains 17:0, 0.1 and 20:1, 1.9%.
10% DEGS on Chromosorb W.
Chowdhury, A. R., Banerji, R., Misra, G. and Nigam, S. K. (1984). JAOCS, **61**, 1023.
All seeds were collected in Lucknow, India except Entada phaseoloides which was obtained from the Andaman and Nicobar Islands.

TABLE 7.50
FATTY ACID COMPOSITION OF MINOR SEED OILS.

	10:0	12:0	14:0	16:0	18:0	18:1	18:2	20:0	22:0
Terminalia panicutata		5·8	6·2	25·5	6·6	16·0	32·4	3·3	4·2
Combretum ovalifolium		0·2	16·4	35·4	2·1	24·1	17·1	2·2	2·5
Dioscorea oppositifolia		0·4	1·3	21·1	4·9	49·4	19·5	1·3	1·7
Dioscorea anguina	0·4		0·3	12·8	5·7	25·8	54·1	0·8	0·5
Stachytarpheta indica			0·5	6·4	5·0	17·9	69·8	0·3	0·1
Aristolochia indica			0·7	18·2	4·3	66·6	4·3	3·3	2·6
Sanseveria cylindica	0·2	0·3	1·7	26·9	6·1	27·2	32·0	3·0	2·6
Sanseveria zeylanica		0·2	1·1	22·3	5·2	34·4	30·2	1·2	5·4
Chrysalidocarpus lutescens	0·8	41·2	26·8	11·8	1·4	10·5	7·5	—	—
Basella alba			2·1	19·4	7·4	46·4	19·7	3·2	1·8

Daulatabad, C. D. and Ankalgi, R. F. (1983). *Fette Seifen Anstrichmittel*, **85**, 404.

TABLE 7.51
OIL CONTENT, IODINE VALUE AND FATTY ACID COMPOSITION OF JAPANESE PLANTS.

	Oil content (%)	Iodine value	12:0	14:0	15:0	16:0	16:1	17:0	18:0	18:1	18:2	18:3	20:0	22:0
Boehmeria longispica Steud.	12·1	149·9	trace	trace	trace	5·2	—	trace	2·3	7·9	82·7	0·9	0·6	0·4
B. spicata Thunb. Urticaceae	10·6	152·2	—	0·2	—	5·2	—	—	1·5	7·8	83·5	1·5	—	0·3
B. nivea Gaud. var. *concolor* Makino	17·5	147·0	—	trace	—	6·2	trace	0·1	2·7	9·4	80·4	1·2	—	—
Tricyrtis affinis Makino	22·0	151·2	trace	0·2	—	5·3	—	trace	1·4	11·4	80·2	1·4	—	0·1
Hosta longipes Matsum.	28·5	152·4	—	trace	—	5·8	0·1	—	1·0	9·5	82·5	0·8	trace	0·3

High linoleic oils have been isolated from achenes and seeds of 5 Japanese plants.
Kato, M. Y. and Tanaka, T. (1981). *JAOCS*, **58**, 866.

7.5 REFERENCES

1. HILDITCH, T.P. (1934). *Chem. & Ind. (Rev)*, **54**, 139, 163 and 184.
2. EGLINTON, G. and HAMILTON, R.J. (1963). *Chemical Plant Taxonomy*. Ed. T. Swain. Academic Press, London.
3. ERDTMAN, H. (1956). *Perspectives in Organic Chemistry*. Ed. A. Todd. Interscience, New York.
4. HILDITCH, T.P. (1956). *The Chemical Constitution of Natural Fats*. 3rd edn. Chapman and Hall, London.
5. SHORLAND, F.B. (1963). *Chemical Plant Taxonomy*. Ed. T. Swain. Academic Press, London.

8

Application of Modification Techniques

J. PODMORE

Pura Foods Ltd, Liverpool, UK.

8.1 INTRODUCTION

Fats are obtained from animal, vegetable or fish sources as shown in Table 8.1 and the most common fatty acids found are C12, C16 or C18 and are either saturated or contain one or more double bonds.

Each fat has a characteristic distribution of fatty acids. The physical, functional and organoleptic properties of these fats are in part a function of that fatty acid composition, but also a function of the fatty acid distribution in the triglyceride comprising such fats. Some examples of typical triglycerides are shown in Fig. 8.1 by way of their carbon number distribution.

Although nature has provided a large number of species consisting of a variety of triglyceride types and mixtures relatively few can be economically harvested in any quantity. Those that are may be lacking in oxidative, hydrolytic, crystallographic, organoleptic or polymorphic stability making them unsuitable or at least undesirable for certain food applications.

The advent of chemical processing techniques, i.e. hydrogenation and interesterification, and physical processing, fractionation, has now allowed the processor the capability of modifying one or more of the properties previously mentioned.[1] It is now possible to produce fats that bear little relation to the natural fat and provides the possibility of providing a 'tailor made' product to suit a particular food. Further it can give the processor a range of alternative raw material sources, thus improving commercial viability.

TABLE 8.1
TYPICAL FATTY ACID COMPOSITIONS OF THE COMMON EDIBLE FATS & OILS[a]

	Iodine value	C_4	C_6	C_8	C_{10}	C_{12}	C_{14}	C_{15}	C_{16}	$C_{16:1}$	C_{17}	C_{18}	$C_{18:1}$	$C_{18:2}$	$C_{18:3}$	C_{20}	C_{22}	$C_{22:1}$	C_{24}
Butterfat	25–42	4	2	1	2	3	13	1.5	26	2	0.5	13	28.5	3	0.5				
Cocoa butter	32–40								26 (23–28)			35 (31–35)	35 (33–39)	3 (2–4)		1			
Coconut oil	7–13		0.5 (5–10)	8 (4–11)	6 (39–54)	49 (13–19)	17 (7–11)		9			2.5 (1–4)	6.5 (5–8)	1.5 (0.5–3)					
Cottonseed oil	99–121						0.9 (0.5–2)		24 (17–29)	0.6 (0.5–2)		2.5 (1–4)	17.5 (13–44)	54 (33–48)	0.5 (0.1–2)				
Groundnut oil	84–102						12		12 (6–16)			3.5 (1–7)	47 (36–72)	31 (13–45)		2.5 (1–3)	3		1
Lard	53–68						2 (0.5–3)		25 (20–32)	3 (2–5)	0.5	13 (5.0–24)	45 (35–62)	10 (3–16)		1.5[b]	(1–5)		
Maize oil	110–128								12 (8–19)			2 (0.5–4)	28 (19–50)	57 (34–62)	1	(2)			
Olive oil	76–90								14 (10–18)	1		2.5 (2–3)	71 (57–78)	10 (6–14)	0.5	1			
Palm oil	45–56						1 (0.5–6)		45 (32–51)	0.1		4.5 (1–8)	39 (34–52)	9.5 (5–12)	0.2	0.2			
Palm kernel oil	14–24			4 (2–6)	4 (3–7)	50 (44–54)	16 (14–19)		8 (6–10)			2.5 (1–4)	13.5 (9–19)	2 (0.5–3)					
Rapeseed oil (high erucic)	97–110								3			1.5 (1–2)	24 (13–30)	15 (10–25)	7.5 (5–10)	1 (1–3)		35 (20–50)	1
Rapeseed oil (low erucic acid)	110–115								4 (2–5)	0.5		2 (1–2)	60	20	10.5	1.5[b]		1.5	
Safflower oil	138–151								6.5 (3–4)			2.5 (1–10)	13 (4–42)	78 (55–81)					
Soyabean oil	125–138								11 (7–12)			4 (2–6)	23 (19–30)	54 (48–58)	8 (4–10)				
Sunflowerseed oil	122–139					0.5	0.2		7 (3–10)			4.5 (2–6)	19 (14–65)	68 (22–75)	0.5	0.3			
Beef tallow	35–51						3.5 (1–6)	1.5	26 (20–37)	3.5 (1–9)	1.5 (0.5–2)	22 (6–40)	39 (26–50)	2.5 (0.5–5)	0.5				

[a] As considerable variations can occur, ranges are given in parentheses.
[b] C 20:1.

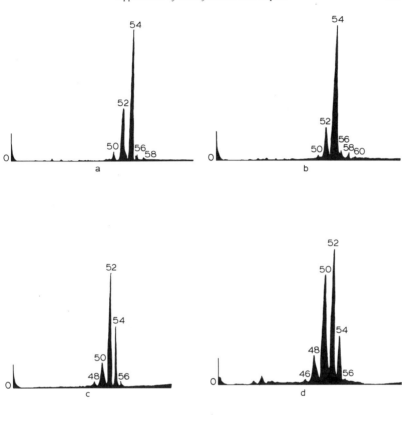
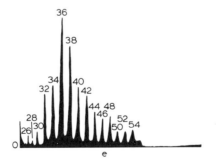

FIG. 8.1. Typical triglyceride structures: (a) soybean oil, (b) sunflower oil, (c) lard, (d) palm oil, (e) palm kernel oil.

The objectives of applying modification techniques are given below and will be illustrated with a number of examples:

1. In order to obtain a fat that meets certain performance characteristics that are not found in nature.
2. Utilisation of cheaper feed stock providing the food processor with a fat of characteristics similar to a more expensive alternative.
3. Improvement in resistance to oxidation. Particularly relevant in hydrogenation.
4. Improvement in palatability—applied to use of stearines.
5. Modification of the crystallisation behaviour of fats.
6. The provision of fats that are nutritionally more acceptable, i.e. reduce saturates and 'trans' fatty acids and increase the polyunsaturated fatty acids.

8.2 SOYBEAN HYDROGENATION

Oils are used very extensively in frying and so as such are not strictly ingredients. However, they are invariably absorbed by the foodstuff and so have an impact on flavour and palatability. They must not only be a good heat transfer medium, they must be bland in flavour, have good resistance to the breakdown processes that are exerted during frying, i.e. hydrolysis, oxidation, polymerisation, discoloration etc.

In 1946[2] the USA had to decide how to use the large quantities of soybean oil that were available. However, it was soon discovered that it gave off a distinctive odour and flavour very quickly when used as a heavy duty frying medium.

The problem was traced to the linolenic acid content (about 7%) which is oxidised much more rapidly than linoleic or oleic acid. Thus by hydrogenation this unsaturation was removed and a stable frying fat could be obtained. However, if the iodine value was reduced too far too much saturated fatty acid was built up with the result that a solid frying medium resulted.

Continued experimental work showed that by using selective catalysts and conditions the iodine value could be reduced from 130 to about 100 without significant build-up of saturated fatty acid, but the linolenic acid could be reduced from 7% to 2% to give a significant improvement to resistance to oxidative breakdown. This technique is often referred to as

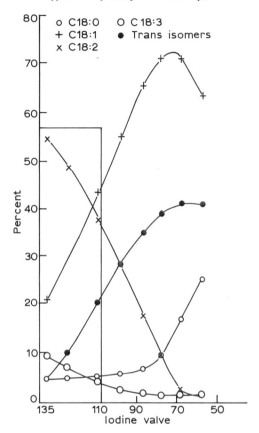

FIG. 8.2. Hydrogenation of soybean oil at 140° and 3 atm H_2 pressure with 0·02% catalyst as Ni.

'brush' or 'touch' hydrogenation and the change in unsaturation is shown in Fig. 8.2.

This product, however, at refrigerator temperatures went cloudy and deposited stearine, so in bottled oil applications it was unacceptable. Thus fractionation in the form of 'winterisation' was introduced to give a completely clear oil with enhanced oxidative stability.

In industrial applications these 'brush' hardened oils have a small percentage of 'stearine' added which creates nuclei and a 'slurry' frying medium is formed and by ensuring small particle size distribution the 'stearine' is kept in suspension.

The approach is now being applied to rapeseed oil which contains about 10% linolenic acid and can develop a 'rubbery' odour and flavour in use.

8.3 LARD—AS A SHORTENING

The classic case for extending the range of uses of a fat by modification is the application of random interesterification to lard.

Lard has always been used extensively as a domestic frying medium as well as a fat for use in making short pastry doughs. However, attempts to use it in a more general purpose shortening showed that its very crystalline texture led to poor creaming (i.e. air incorporation properties).

Examination of the triglyceride structure showed there was a predominating asymmetric triglyceride—OPS with palmitic acid in the β or 2 position. This led to β rather than β' crystal types.[3]

Interesterification randomised the fatty acids and significantly improved the ratio of symmetrical and asymmetric triglycerides. On processing in a scraped surface heat exchanger the product was found to have a finer crystal structure (Fig. 8.3).

The lard was now found to have good creaming properties and could be used in a wide range of bakery applications. Additionally, it could be successfully used as a major ingredient in margarine formulations. This has been extended further by the interesterification of lard and liquid oil blends to give good quality tub margarines with an improved polyunsaturated to saturated ratio.

8.4 MARGARINE DEVELOPMENT

The history of the development of margarine illustrates how important has been the application of modification techniques to the introduction and consequent development of the production of a coarse fat for use in cooking and baking to a spread usable at refrigerator temperatures.

In 1869 Mege Mouries's invention relied on imitating what he perceived as the natural process.[4] Thus he carefully rendered fresh tallow and then slowly crystallised the fat at 25–30°C, i.e. he fractionated it. By pressing this grainy mass he obtained 60% of a soft semi-fluid fraction and 40% of hard white fat. The semi-fluid material was called 'oleo

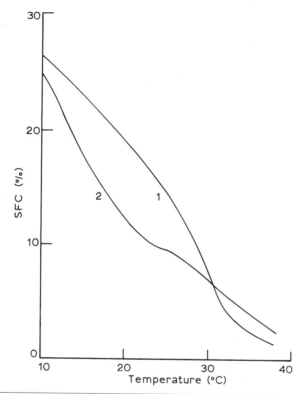

	S_3	SUS	Triglyceride SSU	USU	UUS	U_3
Lard	2	3	20	39	15	18
Interesterified	5	8	17	15	29	26

FIG. 8.3. The effect of interesterification of lard. (1) Lard; (2) interesterified; SFC, Solid Fat Content.

margarine' and the white fat 'oleo stearine'. He wrongly assumed he had butter fat in its purest form because of its physical similarity to natural butter. He felt the shorter chain fatty acids were breakdown products caused by enzymes in the milk.

Using the oleo margarine, skimmed milk and water he formed a 35% fat cream which was churned to a butter-like product. This gave the first margarine which was coarse and grainy in texture. Limitations on the availability of oleo margarine led to developments in the composition of the oil blend. The first step was to carefully prepare beef fat or lard and soften this with cottonseed oil for use in the margarine. Thus blends such as 70% oleo margarine, 10% premier jus and 20% liquid vegetable oil were introduced. Premier jus is the best class of beef fat obtained by low temperature rendering of specific body portions.

The production of beef fat could not keep pace with the demand of the margarine industry. Fortunately, refining techniques introduced at the end of the 19th century opened up the range of vegetable oils that could be purified for use in margarine, these were corn oil, cottonseed oil, coconut oil and palm kernel oil. The cost and availability of oleo margarine continued to limit margarine so more and more palm kernel oil and coconut oil were used, which led to seasonal problems.

In 1903 Normann patented his process for hydrogenation of unsaturated hydrocarbons. By 1912/13 hydrogenated fats appeared on the market and still remain the most important raw material in margarine and shortening. This meant margarines could be more easily adjusted to temperature conditions and also cheaper raw materials could be used, for example fish and whale oil, for not only could it be made harder and bland in flavour, but removal of the unsaturation improved its resistance to oxidation.

These changes, combined with changes in techniques for emulsion processing, changed margarine from a product for cooking and baking into one used as a spread. This led to the requirement of oil blends with good resistance to temperature change combined with good spreadability and lack of waxiness at body temperature. The development of catalysts to produce 'trans' fatty acids was of considerable assistance as it not only improved the melting curve but also assisted in providing good crystallising properties. Blending hydrogenated fats, natural fats and liquid oils led to considerable blend flexibility. The development into refrigerated margarine packed in tubs saw other changes as high levels of liquid oil were required. It was found that by blending high melting components and then interesterifying them, significant changes in melting behaviour could be achieved (Fig. 8.4).

The risk of graininess and post hardening became a serious problem which was found to be most acute in the case of sunflower oil margarine. It was found that due to the very high concentration of C18 fatty acids large coarse crystal agglomerates developed in sunflower/hydrogenated sunflower oil blends leading to 'sandiness'. It was found that if a blend of the hydrogenated sunflower oil component and sunflower oil representing about 60% of the oil blend was interesterified the risk of sandiness was removed.[5]

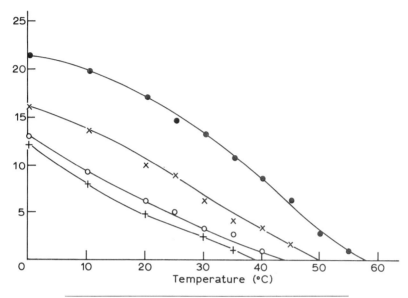

	Interesterified		Blended	MP
	Hardstock hydr. Co/Soy[a]	Soybean oil	soybean oil	(°C)
●	20%	0%	80%	45·1
×	20%	20%	60%	37·7
○	20%	40%	40%	38·1
+	20%	80%	0%	31·0

[a] A cottonseed oil/soybean oil combination.

FIG. 8.4. Margarine fat formulations obtained by hydrogenation and esterification.

Later it was found that by introducing a small quantity of high melting stearine, predominantly C16 fatty acids interesterified with the hydrogenated sunflower oil, the risk of sandiness was also removed![6]

These improvements opened the way for the greater use of the nutritionally attractive oils. Thus high polyunsaturated to saturated ratios and high polyunsaturated fatty acid formulations were available. The presence of 'trans' fatty acids still creates a question nutritionally. However, by the use of fully saturated vegetable fats or the fractions blended with lauric fats and interesterified, this difficulty is successfully removed.[7]

Continued development of the understanding of the contribution of triglyceride structure has led to improvements in blending on the basis of triglyceride. Blending to produce margarine oil blends can now be discussed in terms of triglyceride structure. A patented example for a tub margarine the oil blend should have:

(a) Triglycerides containing three saturated fatty acids (S_3) not greater than 6% as these give a waxy palate sensation.
(b) Triglycerides with a single unsaturated fatty acid (i.e. S_2U) should be greater than 15% as these provide the consistency with mouthfeel.
(c) The $S_2U:S_3$ ratio should be 3.

Thus in the fatty acids distribution there should be:

> C18:0   greater than 15%.
> C18:1   greater than 15%.
> C18:3   greater than 30%.

for example the following proportions lead to quick crystallisation:

> C18:0   20–33%.
> C18:1   20–30%.
> C18:2   40–55%.

This has been achieved by:

(1) Interesterifying a blend of 46% SFO, 14% HSFO 32 and 40% HSFO 69. Fractionating at 34°C to remove tristearate and using olein in margarine.
(2) The blend from (1) could then be blended with 30% SFO.

This example shows the level of sophistication that can be reached in using modification techniques. All three currently available techniques have been used in this example to achieve a stable margarine oil made with sunflower oil, modified to crystallise in a stable form, low in 'trans' fatty acids and saturated fatty acids which affect mouthfeel.

Finally, it is still the margarine manufacturers' aim to make margarine truly like butter and work has been carried out to achieve this end.

Here again in formulating the oil blend, fat modification techniques have been used. In two patents it has been described how a cottonseed oil hydrogenated to 36°C slip melting point was then fractionated and the olein blended with coconut oil, palm and sunflower oil to give a melting curve similar to butter and a plasticity approaching that of butter.[8]

In the second case a tallow olein fraction blended with sunflower oil or rapeseed oil was interesterified and then reblended with a proportion of tallow olein to give a firm texture at room temperature reminiscent of butter.[9]

The examples given I believe indicate that there is now considerable flexibility in selection of oils for margarine manufacture and what is required is an understanding of the triglyceride structure and the application of modification techniques to achieve the relevant triglyceride distribution.

8.5 PALM OIL UTILISATION

Palm oil is a raw material which by the use of all the modification techniques described can be utilised in a very wide range of products. The oil is easy to fractionate, it has good oxidative stability and interesterification significantly modifies its crystallisation behaviour.

Palm is now a regularly available and economically viable oil for the refiner and processor. Additionally, it has been demonstrated how it can be effectively separated into stearines and oleins by fractionation[10] and further modified by interesterification.[11]

Palm oil has long been known as a good heavy duty frying medium because of its relatively low polyunsaturation and the slip melting point low enough to avoid excessive waxiness in most applications. Palm olein is also a very stable frying medium and when lightly hardened easily stands comparison with any other frying fats for resistance to breakdown.

Palm oil was introduced to the blends of margarines and shortenings many years ago, though its use was limited (to about 40% of a blend) because it was found that at higher levels it could cause slow crystallisation leading to graininess and post hardening in the product when processing in the conventional manner.[12]

Examination of the triglyceride structure of palm oil has given the likely reason for this phenomenon. It is suggested that graininess is associated with the symmetrical to asymmetrical triglyceride ratio and the high level of the symmetrical P-O-P triglyceride, 40–45%. This point has been investigated and demonstrated by measuring the percentage solids in solutions of triolein at a range of temperatures when the relative speed of crystallisation is demonstrated.

The increasing demand for all vegetable based products and the ready availability of palm oil has led to finding ways of increasing its use.

Interesterification of palm oil significantly changes the proportion of P-O-P to give a faster crystallising fat (Table 8.2). The drawback is that the S3 proportion is increased leading to a higher slip melting point and shallower melting curve. Hydrogenation also affords a route in that P-E-P is generated. However, S3 is again increased.

TABLE 8.2
PROPERTIES OF PALM OIL BEFORE AND AFTER INTERESTERIFICATION

Triglycerides	Non-interesterified	Interesterified
S_3	7·9	12·4
SUS	42·8	12·2
SSU	6·6	24·2
SU_1	35·7	35·8
U_1	6·8	0·5
Ratio SUS/SSU	6·5	0·5
Solids (%)		
10°	50	50
20°	22	35
30°	7	18
35°	4	13
40°	0·5	9

It can be seen from these remarks that by a judicious use of palm, interesterified palm, palm stearine and palm olein a blend can be made that can be crystallised successfully to give a margarine or shortening capable of performing up to traditional standards in a bakery application.

Palm stearine and palm olein can both be hydrogenated to give melting curves that can be used to produce both successful low melting and high melting puff pastry margarine blends when blended together with proportions of liquid vegetable oils to control the shape of the melting curve and the ultimate plasticity.

By blending palm stearine and, say, rapeseed oil in proportions of 70:30 a successful bread fat can be produced for use in high speed bread production. If the blend is interesterified a very useful cooking fat is produced, though there is the opportunity to add palm olein and palm stearine to adjust melting curve and melting point to give a product acceptable for the duty (Table 8.3).

TABLE 8.3
BREAD FAT AND COOKING FAT BLENDS

	70% Palm St. 30% Soybean oil	Interesterified blend
SFC		
20°C	29	22
30°C	16	11
35°C	12	8
40°C	9	5
Slip point (°C)	44·0	41·0

The use of small quantities of, say, H. Rape or H. Soya significantly improve the properties. However, the palm content can be raised to say 75% without risk of graininess in margarine, shortenings and cooking fats by the use of fractions and interesterification.

The NACNE[13] and COMA[14] reports have both stimulated interest in the nutritional quality of food we eat. More especially in the levels of fat and saturated fat we eat.

There is also a strong case for introducing fats into our diet that contain increased quantities of essential fatty acids (polyunsaturated fatty acid), i.e. linoleic and linolenic acids. An additional constraint is the belief that large quantities of 'trans' fatty acids may be harmful to health in the same way as saturated fats but to a much lesser degree.

Fats with increased polyunsaturated to saturated ratios are now being sought by food processors. As well as a demand for a limitation on the levels of 'trans' fatty acids.

Additionally, there has been a reaction against the use of animal fats such as tallow and lard because of their relatively high saturated fat content.

These requirements have led to a much greater concentration on the use of fractionation, interesterification and hydrogenation techniques that will suppress 'trans' fatty acid development, or hydrogenation to a point where relatively little unsaturation remains.

An example of this requirement for a nutritionally improved oil blend has been the development of a biscuit dough fat. These dough fats have in the past incorporated a wide range of fats, both vegetable and animal based, as well as considerable quantities of hydrogenated fats. In developing alternatives it is essential that there is no loss of oxidative stability nor eating quality, and the risk of 'fat bloom', must not be increased, thus fats likely to develop a coarse crystal structure must be

TABLE 8.4
INTERESTERIFIED DOUGH FAT BLENDS

Blend	90% PO 10% PKO	75% PO 25% PKO	70% PO.St. 30% PKO
SFC			
20°C	29	29	38
30°C	13	11	17
35°C	8	4·5	8
40°C	4	nil	0
SM Pt (°C)	37·0	36·0	37·0

avoided. It can be seen from Table 8.4 that blends incorporating interesterified blends of palm stearine or palm with palm kernel oil will provide a very adequate replacement for the type of products used traditionally without loss of oxidative stability. Thus by using fractionation and interesterifiction a fat can be produced free of 'trans' fatty acids and a marginally better polyunsaturated to saturated ratio.

8.6 SUMMARY

To summarise, the selected examples show that by an understanding of the triglyceride structure of fats and the changes created as a result of the major modification processes currently available, it is possible to produce blends of fats from an increasingly wide range of oil sources.

The food industry has developed enormously in sophistication under the pressure of demands of economics, pre-prepared and convenience foods, which in turn places demands on the ingredient supplier to meet changes in technological needs and cost constraints. These changes will continue to test the ingenuity of the oils and fats processor to produce fats to meet those changes.

8.7 REFERENCES

1. THOMAS, E. A. and PAULICKA, F. R. (1976). Chem & Ind., 774–9.
2. DUTTON, H. J. (1981). *JAOCS*, No. 3, 234–6.
3 FAUR, L. (1977). *Revue Francaise Corps Gras*, **24** (No. 2), 85–91.
4. VAN STYVENBERG, J. H. (1969). *Margarine*. Liverpool University Press.
5. Canadian Patent, 1,166,892.

6. Canadian Patent, 1,185,117.
7. U.S. Patent, 4,341,813.
8. European Patent Application—Publication No. 0109 721.
9. Canadian Patent, 1,194,360.
10. DEFFENSE, E. (1985). *JAOCS*, **62** (2), 376–84.
11. LANING, S. J. (1985). *JAOCS*, **62** (2), 400–7.
12. DUNS, M. L. (1985). *JAOCS*, **62** (2), 408–10.
13. Nutritional Guidelines for Health Education in Britain (1983) National Advisory Council on Nutritional Education. Published by the Health Education Council.
14. DHSS. (1969). *Diet and Cardiovascular Disease*. Report of the Advisory Panel on Medical Aspects of Food Policy, HMSO, London.

Index

Acacia fornesiana, 147
Acrocomia sclerocarpa, 145
Actinodaphne angustifolia, 144
Actinodaphne hookerii, 144
Acyl groups
  attached at each of three *sn* positions, 18–22
  attached to carbons 1 and 3, 16–18
    chemical methods, 16–17
    enzymic method, 17–18
    physical methods, 16
  sn 1- and *sn* 3-positions, 20
  sn 2- and *sn* 1-positions, 20
Albizzia lebbek Benth., 163
Albizzia lucida Benth., 163
Albizzia richardiana King and Prain, 163
Aleurone lipids, chemical markers for, 90–4
Aliphatic aldehydes, 33
Allanblackia floribunda, 144, 149
Allanblackia parviflora, 144
Allanblackia stuhlmannii, 144
Almond, 118, 139
α modification, 4
Amaranthus tricolor, 147
Androgaphis paniculata, 154
Anethium graveolens, 151
Anisidine test, 33
Antioxidants, 38
Apocyanaceae, 143
Aristolochia indica, 164
Arrhenius Activation Energy, 31
Asclepias incarnata, 151

Asclepias syriaca, 151
Autoclaves
  design, 47–52
  pressure levels, 43
Avena sativa, 117
Azadirachta indica, 160

Basella alba, 164
Bauhinia species, 152
β modification, 2
Bixa orellana, 147
Bloom, 8
Boehmeria longispica Steud., 165
Boehmeria nivea Gaud. var. *concolor* Makino, 165
Boehmeria spicata Thumb. Urticaceae, 165
Bombax malabarica, 145
Brassica juncea, 110
Brassicaceae, 156, 158
Bread-making process, 79–107
  correlations between baking quality and lipid composition of wheat grain and flour, 99
  interfacial activity, 100–1
  lipid association with proteins, 104–5
  liquid–crystalline phases, 102–4
  mechanisms by which lipids influence baking quality, 99–105
  observed effects of flour lipids in, 97–9
Butter
  fishy taints, 38

Butter—contd.
  photo-oxidation, 38
  rancidity, 36
Buttermilk, lipid oxidation, 37
Butyrospermum parkii, 144

Calliopsis elegans, 154
Canarium commune, 145
Canarium ovatum, 145
Capparidaceae, 143
Carapa procera, 160
Carum carvi I, 151
Carum carvi II, 151
Caryocar villosum, 145
Cassava cultivars, 124
Cassia auriculata, 147
Cedrella odorata, 160
Celosia cristata, 147
Celosia pyramidalis, 162
Centrosema arenarium, 161
Centrosema laurifolia, 161
Centrosema pubescens, 161
Centrosema sp., 161
Centrosema virginionum, 161
Chrysalidocarpus lutescens, 164
Cinnamomum camphora, 144
Cleome viscosa, 143
Clitoria ternatea, 161
Coberine, 8
Cocoabutter (CB), 8
Cocoabutter equivalent (CBE), 8
Cocos nucifera, 145
Combretum ovalifolium, 164
Compatibility of fats, 8–9
Compound formation, 7, 10
Convolvulaceae, 153
Convolvulus arvensis, 153
Copper additions, 38
Corchorus capsularis, 143
Corchosus acutangulus, 143
Coriandrum sativum, 151
Corochorus trilocularis, 154
Cottonseed varieties, 112–13
Crotalaria heyneana, 154
Cryptostegia grandiflora, 162
Crystallization behaviour, 2–5
Crystallization processes, 4–5

Cuminum cyminum, 151
Cyclopropane containing seed oils, 150
Cyperaceae, 143
Cyperus iria, 143
Cyperus rotundus, 143

Dacryodes rostrata, 145
Desplatzia dewevrei, 149
Differential Scanning Calorimetry (DSC), 4–5
Diglyceride kinase, 20–2, 28–9
2,3-dioleyl-2-palmitoylglycerol (OPO), 8
Dioscorea anguina, 164
Dioscorea oppositifolia, 164
1,3-dipalmitoyl-2-oleylglycerol (POP), 8
1,3-dipalmitoyl-2-oleylglycerol/1, 2-dipalmitoyl-3-oleylglycerol (POP/PPO), 8–9
1,3-distearoyl-2-oleylglycerol (StOSt), 8
Dough-mixing, lipid–protein associations during, 96
Duranta repens, 147
Durio silbethinus, 138

Edible oils, imports of, 34–5
Elaeis quineensis, 145
Emblica officinalis, 154
Endosperm non-starch lipid composition of wheat grain, 94
Energy saving, hydrogenation, 43
Entada phaseoloides Merrill, 163
Entada scandens Benth., 163
Enthandrophragma angolense, 160
Erisma calcaratum, 145
Eutectic behaviour, 7
Evolvalus alsinoides, 153

Fatty acid compositions, 109–66
Fatty acid sequence, 13–30
  methods for determining, 15–22
Fatty acids
  milk fats, 61–6

Fatty acids—contd.
  milk triacylglycerols, 66–8
  positional distribution in triglycerides of African peanut varieties, 22–3
  stored flour, 96
Fig seed, 129
Flame-ionisation detector, 60
Flour, lipid composition of, relationship with lipid composition of wheat grain, 86–94
Flour milling process, 87–9
  lipid redistribution during, 89–90
Flour oil content and colour grade, 89–90
Foeniculum vulgare, 151

Garcinia indica, 144
Garcinia kola, 149
Garcinia morella, 144
Gas chromatography (GC), 14
Gas chromatography–mass spectrometry (GC–MS), 14
Gas–liquid chromatography (GLC), 57, 59–60, 62–5, 69
Glycerol, 14
Gnetum scandens, 144
Gomphrena globosa, 162
Graininess in margarines, 8
Grape seed, 129
Grignard reaction, 18–22, 24–6, 26–8
Groundnut, 129
Gynandropsis pentaphylla, 143

Hempseed, 129
Hevea brasiliensis, 130
Hibiscus species, 150
High-performance liquid chromatography (HPLC), 14, 57–60, 63, 65, 67, 69–73
High-temperature gas–liquid chromatography (GLC), 73
Hildebrand equation, 7
Hildebrand solubility curves, 8
Hodgsonia capniocarpa, 145
Holmskioldia sanguinea, 147

Horiuti–Polanyi mechanism, 53–4
Hosta longipes Matsum, 165
Hydrogenation, 41–56
  autoclave design, 47–52
  classes of oils processed, 46
  dead end technique, 45
  energy saving, 43
  energy/time saving modification, 44
  equipment layout, 41–7
  hydrogen leak precautions, 41–2
  hydrogen purity levels, 46
  hydrogen storage, 46
  layout of hardening plants, 45
  phases involved in, 41
  pressure levels, 43
  production planning, 45
  reactions at the catalyst surface, 52–5
  soybean, 170–1
  sunflower oil, 174
Hydroperoxides, 33
Hymenantherum tenuifolium, 147

Inga bigemina Hook and Arn., 163
Ipomoea biloba, 153
Ipomoea hederacea, 153
Ipomoea involucrata, 147
Ipomoea pestigridis, 147
Ipomoea quamoclit, 153
Ipomoea sepiaria, 147
Ipomoea tuberosa, 153
Irvingia barteri, 145
Irvingia gabonensis, 145
Irvingia oliveri, 145
Iso-dilatation curves, 6, 10
Iso-solids diagrams, 6, 8
Isotoma longiflora, 162

Jasminium officinale, 162
Jatropha gossypifolia, 162

Kernel oils, 149
Khaya senegalensis, 160

Lantana sellowiana, 147

Lard, 169, 172
  stereospecific analysis of, 23
Laurelberry, 129
Laurus nobilis, 145
Leguminous seeds, 163
Leucaena leucocephala (Lamk.) Wit., 163
Leucas cephalotes, 151
Leucas linifolia, 162
Linoleic acid, 23, 131
Linseed, 129
Linum seed oils, 141–142
Lipid composition
  bovine milk, 59
  flour, relationship with lipid composition of wheat grain, 86–94
  wheat grain, 82–6
    relationship with flour lipid composition, 86–94
  white flours during prolonged storage, 95
Lipid degrading enzymes in wheat grain, 85–6
Lipid hydroperoxides, 39
Lipid oxidation, 31–9
  accelerated, 31
  avoidance of, 37
  buttermilk, 37
  facile nature of, 33
  flavours, 32
  initiation, 32, 35
  intervention butter oil, 36
  liquid milk, 37
  meat, 32
  secondary oxidation products of, 33
  susceptibility to, 37
  transported oils, 35
Lipid protein associations during dough-mixing, 96
Lipids
  analysis, 57–78
  association with protein, 104–5
  class separations, 58–61
  distribution in wheat grain tissues, 82–5
  redistribution during milling, 89–90
  role in baking of bread, 96–105

Lipoxygenases, 32
Litsea cubeba, 144
Litsea longifolia, 144
Litsea sebifera, 145
Litsea zeylanica, 144
Lochnera pusilla, 143
Loop hydrogenation reactor, 51–2
Lophira alata, 145
Lophira procera, 145
Lovoa trichilloides, 160
Lucuma species, 155

Madhuca butyracea, 144
Madhuca latifolia, 144
Madhuca longifolia, 144
Maize, 115–16
Mangifera indica, 144, 154
Mango fats, 111
Manihot esculenta Crantz, 124
D-Mannitol, 24
Margarine, development history, 173–7
Melting behaviour, 2–5
Methyl linoleate, 32
Milk, lipid oxidation, 37
Milk butylesters, 63
Milk fats
  fatty acids, 61–6
  lipid analysis, 57–78
Milk triacylglycerols
  fatty acids in, 66–8
  molecular species of, 68–73
Milletia laurentii, 149
Minusops heckelii, 144
Minusops njave, 144
Modification techniques, 167–81
Modification types, 2
Molar heats of fusion, 5
Mowrah fat, 128
Mucuna species, 155
Myrica cerifera, 145
Myrica cordifolia, 145
Myristica fragrans, 145
Myristica malabarica, 145

Nephelium lappaceum, 145

Nephelium mutabile, 145
Noisette, 139
Nut fatty acid composition, 139

Oat strains, 117
Octadecenoic acid isomers, 65
Oleic acid, 23, 55, 131
Opuntia dillenii, 162

Palm kernel oil, 169
Palm oil, 169
  products, 167
  properties before and after interesterification, 178
  utilisation, 177–80
Palm varieties, 127, 136–7
Palmitic acid, 23
Paloquium oblongifolium, 144
Pancreatic lipase, 20–2, 28
Panda oleosa, 149
Paraglycine hedyseroides, 161
Paraglycine radicosa, 161
Parkia biglandulosa W. and A., 163
Partition chromatography, 57
Pavetta indica, 162
Peach kernel, 129
Peach varieties, 118
Peanut, 132, 140
  varieties, 120–4, 126
Pentaclethra macrophylla, 149
Pentadesma butyracea, 144
Pergularia daemia, 154
Peroxide Value, 33, 36
Petroselinum sativum, 151
Phase diagrams, 6–8
Phosphatidylethanolamine, 38
Phospholipase A, 18, 22, 24–6
Phospholipase C, 20, 26–8
Phospholipids, 38
Phosphorylation, 18, 20, 22, 24–9
Phyalis maxima, 147
Physical properties, 1–11
Pimpinella anisum, 151
Pithecolobium bigemina (L) Mart., 163
Pithecolobium dulce Benth., 163
Platonia insignis, 144

Polymorphism, 2–4
  methods of study, 4
Polyunsaturated acyl groups, 38
Polyunsaturated fatty acids, 31, 38, 63
Poppyseed, 129
Product defects, 8–9
Proteins, lipid association with, 104–5
Psychotria dalzellii, 162
Pulse-NMR, 5–6
Pycnanthus kombo, 145

Radial flow flat blade turbine stirrer, 48
Rancidity, 35–37
Rapeseed, 127
Rapeseed mutants, 119
Reversed-phase separation, 63
  cows' milk triacylglycerols, 70–1
  rat milk triacylglycerols, 72
Rhus succedanea, 144
Rubber seed oil, 130

Safflower species, 125
Salvadora oleoides, 145
Salvadora persica, 145
Salvia farinacea, 147
Sanseveria cylindica, 164
Sanseveria zeylanica, 164
Sapium sebiferum, 145
Sesame seed, 129
Shorea rolusta, 144
Shorea stenoptera, 144
Sida humilis, 162
Sida spinosa, 147
Solid fat index (SFI), 5
Solids content, 6–8
Sopubia dulphinifolia, 154
Soya varieties, 125
Soyabean varieties, 114, 131
Soyabeans, 161
Soybean hydrogenation, 170–1
Soybean oil, 169
Spanish oil scandal, 36
Sphingolipids of cows' milk, 66
Stachytarpheta indica, 164
Stearic acid, 23

1-stearoyl-2-oleyl-3-palmitoylglycerol (StOP), 8
Stereoisomeric forms, 14
Stereospecific hydrolysis, 23–4, 26–8
Stereospecific numbering, 14
Sterol compositions, 118, 131
Stizolobium species, 155
Sub-α modification, 4
Sunflower oil, 169
  heating at various temperatures in air, 23–4
  hydrogenation, 174
  overheating, 35
Sunflower seed, 131, 133–5
Syrian Scabious, 129

Tea seed, 129
Teramnus uncinatus, 161
Teramnus volubilis, 161
Terminalia catappa, 160
Terminalia glucausens, 160
Terminalia panicutata, 164
Terminalia superba, 160
Theobroma cacao, 144
Thin-layer chromatography (TLC), 14, 57–60, 64–5, 67–8, 73
Tiliaceae, 143
Tobacco seed, 129
Tocopherol
  distribution among wheat grain tissues, 92
  markers for germ and aleurone lipids, 90–4
Toddalia asiatica, 162
Tomato seed, 129
Totox determination, 33
Toxic oxidation products, 36
Treculia africana, 149
Trichilia emetica, 145

Tricyrtis affinis Makino, 165
Triglycerides, correlation of configuration, 24

UV detection, 59, 60, 64

Vateria indica, 144
Venturi mixing jet, 50
Vincetoxicum nigium, 151
Virola bicuhyba, 145
Virola otoba, 145
Virola surinamensis, 145

Wheat flour, storage stability of, 95–6
Wheat germs, chemcial markers for lipids, 90–4
Wheat grain
  endosperm non-starch lipid composition of, 94
  lipid composition, 82–6
    relationship with flour lipid composition, 86–94
  lipid distribution, 82–5
  lipid-degrading enzymes, 85–6
  lipids, 79–107
  median longitudinal section, 80–1
  moisture content, 94
  structure, 79–81
White corn hybrids, 115
Wilbuschewitsch Mixing Jet, 48

X-ray diffraction, 4–5

Yellow corn hybrids, 115
Yellow-seeded rai, 110